Fiber-Reinforced Polymer Composites for Building and Bridge Applications

Fiber-Reinforced Polymer Composites for Building and Bridge Applications

Guest Editors

Xingxing Zou
Lili Hu
Zhiqiang Dong

MDPI

Basel • Beijing • Wuhan • Barcelona • Belgrade • Novi Sad • Cluj • Manchester

Guest Editors
Xingxing Zou
Elite Engineering School
Changsha University of
Science and Technology
Changsha
China

Lili Hu
School of Future Cities
University of Science and
Technology Beijing
Beijing
China

Zhiqiang Dong
School of Civil Engineering
Southeast University
Nanjing
China

Editorial Office
MDPI AG
Grosspeteranlage 5
4052 Basel, Switzerland

This is a reprint of the Special Issue, published open access by the journal *Polymers* (ISSN 2073-4360), freely accessible at: https://www.mdpi.com/journal/polymers/special_issues/Fiber_Rein.

For citation purposes, cite each article independently as indicated on the article page online and as indicated below:

Lastname, A.A.; Lastname, B.B. Article Title. *Journal Name* **Year**, *Volume Number*, Page Range.

ISBN 978-3-7258-4939-0 (Hbk)
ISBN 978-3-7258-4940-6 (PDF)
https://doi.org/10.3390/books978-3-7258-4940-6

Contents

About the Editors

Xingxing Zou

Xingxing Zou is a faculty member at Changsha University of Science and Technology. He obtained his PhD degree at Missouri University of Science and Technology, US. His research interests include non-destructive evaluation (NDE) of structures, and strengthening of structures using fiber-reinforced polymers. He has published more than 60 peer-reviewed articles, with a total citation number on Google Scholar of 1200 with an h-index of 24. He has received nine research grants from organizations including national science foundation of China.

Lili Hu

Lili Hu is a faculty member at University of Science and Technology Beijing. Her research interests include high-performance FRP composite structures and advanced construction technology. She has authored/co-authored over 40 international SCI journal papers leading to over 1000 citations. She is an IIFC council member. She achieved the first engineering application of strengthening a 5G high-rise signal tower. Her research contributions have been recognized by a number of honors, including National Youth Support Talent.

Zhiqiang Dong

Zhiqiang Dong is a faculty member at Southeast University. He obtained his PhD degree at Southeast University, China. He was primarily involved in researching and applying high-performance civil engineering materials, such as the green and low-carbon fiber-reinforced polymer (FRP), iron-based shape memory alloy (Fe-SMA), and ultra-high performance concrete (UHPC) in the fields of active maintenance of infrastructures. He has published more than 100 peer-reviewed papers, including 91 SCI journal papers and 10 EI papers, and the total number of citations on Google Scholar exceeds 3650 with an h-index of 32. He has received funding for six research projects from the Chinese government and also received financial support from eight prestigious enterprises.

Preface

The field of civil engineering has been profoundly enriched by the application of advanced materials such as fiber-reinforced polymer (FRP) composites. These materials, with their remarkable strength, lightweight properties, and resilience in demanding environments, have opened up new possibilities for strengthening existing structures and constructing innovative, sustainable infrastructure. As research continues to evolve, this reprint brings together the contributions of a dedicated group of researchers who participated in the Special Issue on FRP composites. The collected works reflect a shared commitment to deepening our understanding of FRP materials through rigorous experimentation, analytical modeling, and numerical simulation, always with an eye toward practical, real-world applications.

We are deeply grateful to the editors of the journal Polymers, and the authors of this Special Issue. The nine articles featured here explore a range of critical topics, from FRP-strengthened concrete to the long-term performance of FRP under environmental stressors. Each study represents a step forward in our collective effort to ensure that these materials are used safely, efficiently, and responsibly.

We would like to express our sincere appreciation to all the authors, reviewers, and editorial teams whose hard work and thoughtful contributions made this Special Issue possible. It is our hope that this reprint will serve as a modest but meaningful resource for those working to advance the field, and that it inspires further collaboration and innovation in the years to come.

Xingxing Zou, Lili Hu, and Zhiqiang Dong
Guest Editors

polymers

MDPI

Article

Static and Dynamic Performance of Long-Span Suspension Bridges with Flexible CFRP Central Buckles

Maoqiang Wang [1], Taike Zhang [2,3], Huaimao Yang [1], Yaoyu Zhu [1], Bin Liu [4] and Yue Liu [4,*]

1 CCCC Highway Bridges National Engineering Research Centre, Ltd., Beijing 100120, China; wangmaoqiang@bnerc.com (M.W.); yanghuaimao@bnerc.com (H.Y.); zhuyaoyu@bnerc.com (Y.Z.)
2 Guangdong Provincial Highway Construction Co., Ltd., Guangzhou 510101, China; gdbatcz@163.com
3 Guangdong Bay Area Transportation Construction Investment Co., Ltd., Guangzhou 511458, China
4 Research Institute of Urbanization and Urban Safety, School of Future Cities, University of Science and Technology Beijing, Beijing 100083, China; d202410038@xs.ustb.edu.cn
* Correspondence: yueliu@ustb.edu.cn

Abstract

The central buckle is essential for maintaining longitudinal stability in suspension bridges. However, conventional steel buckles are often excessively stiff, leading to stress concentration and insufficient durability. Moreover, they tend to perform poorly under fatigue loading conditions. This study proposes a novel flexible central buckle system based on a Carbon Fiber-Reinforced Polymer (CFRP) to address these limitations. This study proposes a novel flexible central buckle system based on Carbon Fiber-Reinforced Polymer (CFRP) to address these limitations. Taking the long-span Shiziyang Suspension Bridge as a case study, a finite element model is developed to investigate the effects of CFRP central buckles with eight different stiffness levels on the static and dynamic responses of the bridge. The results indicate that a CFRP central buckle with a low elastic modulus achieves comparable displacement control performance to that of traditional steel buckles, while inducing significantly lower internal forces, demonstrating strong potential as a substitute. Based on this finding, a coordinated control strategy combining the CFRP central buckle with end-span restraining devices is proposed. This integrated system reduces midspan displacement and central buckle internal force by 61.1% and 49.8%, respectively. Considering both performance and cost-efficiency, a low-modulus CFRP material such as T300 is recommended. The proposed approach offers a new and effective solution for longitudinal control in ultra-long-span suspension bridges.

Keywords: suspension bridge; central buckle; CFRP cable; static performance; dynamic performance

1. Introduction

Suspension bridges, as a core structural form for long-span applications, have become the preferred solution for ultra-long crossings such as sea straits and deep valleys due to their exceptional spanning capabilities [1–3]. In current engineering practice, semi-floating and fully floating systems are commonly adopted to release longitudinal constraints. However, under vehicular loads, temperature variations, and seismic actions, these systems tend to experience significant longitudinal displacements, which can exacerbate fatigue damage in expansion joints and bearings, thereby compromising long-term durability and driving safety [4,5]. To enhance longitudinal stiffness and global stability, the central buckle has been widely employed as a key component in suspension bridges [6–8].

 https://doi.org/10.3390/polym17131807

At present, mainstream central buckle configurations mainly include rigid truss types and flexible cable types: the former connects the cable and girder through rigid truss members, forming a geometrically incompatible constraint system; the latter introduces flexible longitudinal restraint via prestressed high-strength cables [9–12]. Existing studies have demonstrated that flexible central buckles perform well in improving longitudinal stiffness, reducing tower base moments, and enhancing aerodynamic stability [13,14]. Furthermore, central buckles help reduce torsional displacements of the main girder, contributing to improved deck driving stability [15]. However, their influence on lateral stiffness and deflection control remains limited [7,16]. Their mechanical effects are mainly confined to the longitudinal direction. It is noteworthy that while flexible central buckles help alleviate stress concentration, excessive stiffness can instead lead to internal force amplification, highlighting the importance of a stiffness–ductility balanced design [7]. In response to the complex structural behavior of long-span bridges under non-uniform excitations such as earthquakes, scholars have proposed various models—such as modified traveling wave effects and non-classical damping theories—to better capture system dynamics [17–20]. These studies have revealed that the modal strain energy distribution within the central buckle–cable–tower coupled system significantly influences the global response characteristics. Nevertheless, under antisymmetric mode excitation, geometric deformation of the central buckle and torsional warping of the stiffening girder may diminish its longitudinal control effectiveness, necessitating optimization of axial stiffness and structural configuration.

Despite the central buckle playing a pivotal role in structural regulation, its performance under extreme dynamic loads remains a significant challenge. Rigid central buckles, due to their high stiffness constraints, are prone to stress concentration and have limited adaptability under non-uniform excitations such as earthquakes [7]. In contrast, traditional flexible central buckles—while effective in mitigating stress concentrations—are susceptible to corrosion and fretting wear, leading to reduced fatigue life and difficulty in meeting the elevated safety and durability demands of ultra-long-span bridges.

To address these issues, this study proposes a novel flexible central buckle system based on a Carbon Fiber-Reinforced Polymer (CFRP) as a substitute for conventional steel central buckles. The approach aims to leverage the inherent advantages of CFRP, including its lightweight characteristics, high strength, corrosion resistance, and excellent fatigue performance [21–25]. In particular, CFRP features a low coefficient of thermal expansion and complete immunity to electrochemical corrosion, which makes it especially suitable for applications in coastal or salt-laden environments A finite element model is developed based on a representative suspension bridge with a 2180 m main span and steel truss girder to systematically investigate the impact of CFRP central buckles with varying stiffness on the static and dynamic performance of the bridge. This study focuses on evaluating the influence of CFRP central buckles on longitudinal displacement control, internal force redistribution, and structural wind and seismic resistance. Furthermore, a performance–cost integrated evaluation framework is established, and practical recommendations for component selection are proposed. This provides theoretical support and design reference for the application of FRP materials in critical bridge components.

2. Engineering Project

2.1. Bridge Location and Environment

The Shiziyang Bridge is located in the core area of the Pearl River Estuary in Guangdong Province, China. Figure 1 illustrates a conceptual rendering of the bridge.

Figure 1. Rendering of a suspension bridge.

This location sits at the confluence of saltwater and freshwater, resulting in a highly corrosive environment. The subsurface geological conditions are dominated by thick silt layers and weak foundations, with bedrock buried at considerable depth. In terms of seismic safety, the region falls under Seismic Intensity Zone VII, requiring rigorous seismic design measures to ensure structural integrity. Climatically, the site is subject to a subtropical monsoon climate with abundant annual rainfall, frequent thunderstorms, and regular typhoon activity—imposing stringent requirements on the bridge's wind resistance, drainage systems, and lightning protection design.

2.2. Bridge Design

The Shiziyang Bridge adopts a single-span suspension system. As shown in Figure 2, the span arrangement is 672 m + 2180 m + 710 m, with a total length of 3562 m. The main cable has a span-to-sag ratio of 1:9. Figure 3 shows the visual rendering of the bridge. The main girder adopts a double-deck steel truss structure with an overall deck width of approximately 45 m and a truss depth of 13.5 m. A total of 16 traffic lanes are arranged across the upper and lower decks. The main towers are designed as portal-type steel–concrete composite structures.

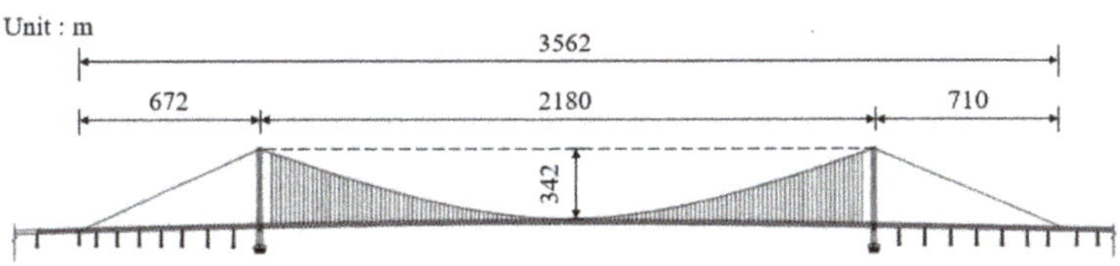

Figure 2. Details of suspension bridges.

Figure 3. Truss girder.

3. Computational Model

3.1. FE Model

A three-dimensional finite element model of the suspension bridge was established using SAP2000 (Figure 4). The modeling process accounted for the geometric nonlinearity of the towers, main cables, and hangers. Initial stress states induced by self-weight and secondary dead loads were introduced to more accurately simulate their effects on seismic response. The stiffening girder, towers, and piers were modeled using beam elements. The main cables and hangers were modeled using truss elements. Rigid arms were added between the lower ends of the hangers and the girder to transfer constraints. The central buckle was modeled using tension-only truss elements, with compressive axial stiffness set to zero to accurately reflect its unidirectional force-transmitting behavior [6]. The cross-sectional area was specified as 25,000 mm^2. The finite element model assumed that the CFRP components and the connecting parts were fully bonded without slip.

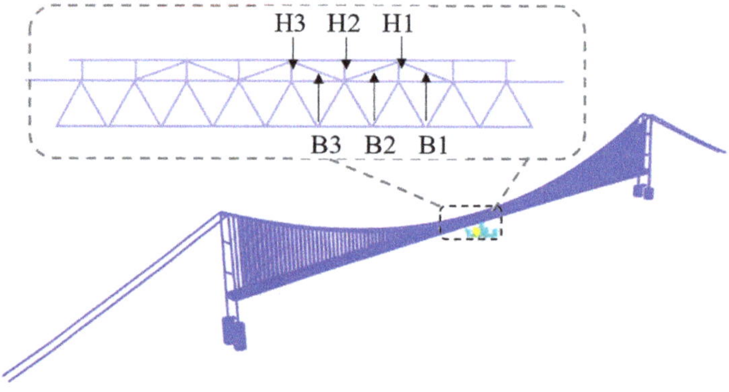

Figure 4. Three-dimensional finite element model of the suspension bridge.

In the initial structural configuration, no girder-end restraining devices were installed at the main towers; support was provided solely through transverse and vertical bearings. To evaluate the mechanical performance of the flexible central buckle system under varying boundary constraints, this study introduces a combined constraint condition incorporating restraining blocks and viscous dampers, simulating the collaborative control mechanism between the girder-end devices and the flexible central buckle. The specific setup is as follows: two restraining blocks are installed at each main tower, totaling four across the bridge, with a restraining gap of 1.1 m. Additionally, four viscous dampers are installed per tower, resulting in eight dampers for the entire bridge. Each damper is assigned a damping coefficient of 3000 kN/(m/s)^ξ, and a velocity exponent of 0.1. This parameter selection was based on previous seismic studies that identified this combination as providing optimal damping performance.

Dampers were modeled using exponential-type Maxwell elements to capture their nonlinear viscous characteristics [26]. End-of-girder limit blocks were simulated with multi-stage elastic link elements: zero stiffness was applied within a displacement range of $\pm$1.1 m, and extremely high stiffness was applied beyond this threshold to simulate rigid collision effects. The pile–soil interaction was represented using equivalent soil springs, with lateral stiffness determined based on the "m" method. Lateral and vertical supports of the main girder were modeled using elastic link elements, and relevant parameters were selected based on the geotechnical investigation report and the *Code for Design of Highway Bridge and Culvert Foundations* (JTG 3363—2019) [27] to ensure reasonable boundary conditions.

The elastic modulus is important for the performance of the central buckle as it directly influences both the static and dynamic behavior of the suspension bridge [7,10]. Based on publicly available material data from Toray Industries, Japan (Table 1), eight types of Carbon Fiber-Reinforced Polymer (CFRP) cables with different elastic moduli were selected as candidate materials for the central buckle: 161 GPa, 168 GPa, 200 GPa, 227 GPa, 264 GPa, 305 GPa, 378 GPa, and 412 GPa. A conventional steel cable with an elastic modulus of 206 GPa was used as the reference for comparison. These materials have been widely adopted in structural engineering applications, covering the commonly used CFRP products ranging from low to ultra-high modulus grades. The impact of different central buckle stiffnesses on the static and dynamic responses of the suspension bridge was systematically evaluated. Three pairs of central buckles were symmetrically arranged at the midspan of the main span, as shown in Figure 5. This study includes two main parts: (1) a comparison of the bridge's static and dynamic behavior under the action of central buckles with different stiffness levels; (2) an analysis of the structural response when the central buckle is used in combination with end-girder restraining devices.

Table 1. CFRP strand types and properties.

Type	T300	T700 G	T800 S	T1000 S	M40 J	M46 J	M55 J	M60 J
E (GPa)	161	168	200	227	264	305	378	412

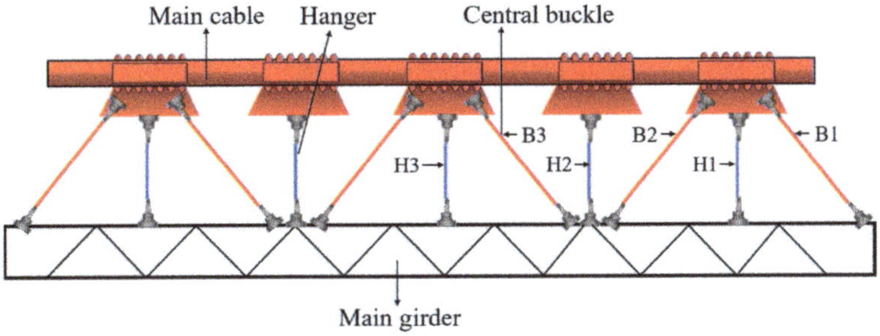

Figure 5. Configuration and numbering scheme of central buckles.

3.2. Load Conditions

The calculation of permanent loads comprehensively considered the self-weight of the main girder, crash barriers, and pavement layers to accurately reflect the actual service conditions. As shown in Figure 6, live loads were applied in accordance with the Class I highway loading standard specified in the *Technical Standard of Highway Engineering* (JTG B01-2014) [28]. Regarding wind loads, long-term meteorological observations at the bridge site were used to determine a basic design wind speed of 37.9 m/s. Under combined wind–vehicle loading conditions, a reduced turbulence intensity was considered, resulting in an effective design wind speed of 25 m/s at deck height. Temperature loads were applied using a uniform temperature rise-and-fall method, with values of +31 °C for heating and −23 °C for cooling scenarios.

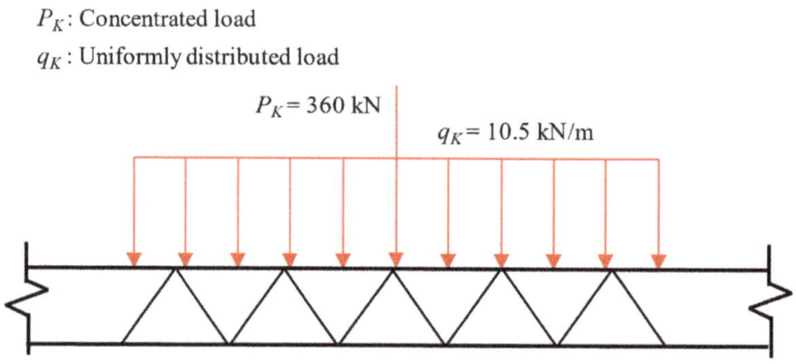

Figure 6. Live loads.

Seismic action was represented by five horizontal ground motion time history curves (Figure 7) provided in the site-specific seismic safety assessment report, corresponding to a 4% exceedance probability in 100 years. Vertical seismic input was set as 0.65 times the peak horizontal ground acceleration, in accordance with the code requirements. To enhance the statistical reliability of the results, the structural response was taken as the mean value of the responses obtained from the five ground motion records.

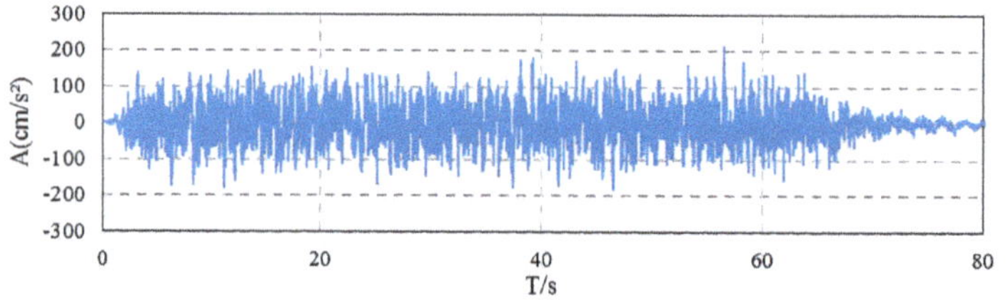

Figure 7. Acceleration time history of ground motion.

4. Results and Discussion

4.1. Static Response Analysis

4.1.1. Girder-End Displacement

To ensure the validity of the finite element model, a multi-level verification approach was adopted. First, the main cable profile under dead load was compared with the theoretical catenary solution, showing good agreement. Second, the modeling methods and parameters for the restraining devices were selected based on established references, software guidelines, and relevant design codes to ensure accuracy and engineering reliability. As a critical control parameter in long-span suspension bridges, the longitudinal displacement at the girder ends directly affects the design of expansion joints and the selection of bearing systems [2,4]. Figures 8 and 9 illustrate the influence of central buckle stiffness on the longitudinal displacement of the main girder. The results for the steel cable central buckle are highlighted in darker shades. To better emphasize the effects of different CFRP cable stiffnesses on structural performance, the figures have been zoomed in and the baseline case (without a central buckle) has been omitted. Under the baseline condition, the end displacement of the girder exhibits a pronounced bidirectional asymmetry: the maximum positive displacement δ_{max} is 2.249 m, and the maximum negative displacement δ_{min} is −2.270 m. The displacement amplitude difference between the cable and the girder, $\delta_{cable} - \delta_{girder}$, is 0.021 m (Table 2), which confirms the geometric nonlinearity associated with large displacement responses in long-span suspension bridges.

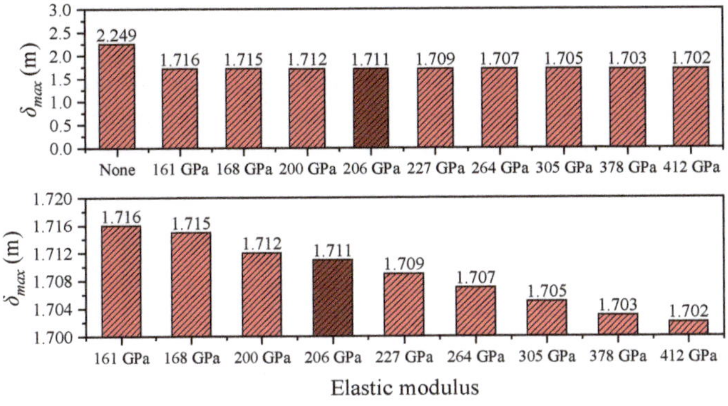

Figure 8. Maximum girder-end displacement under different central buckle stiffnesses in static conditions.

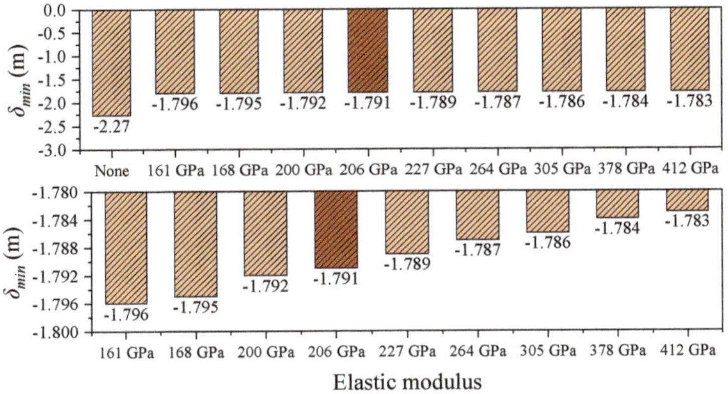

Figure 9. Minimum girder-end displacement under different central buckle stiffnesses in static conditions.

Table 2. Static displacement responses of the structure.

E	δ_{max} (m)	δ_{min} (m)	$\delta_{cable} - \delta_{girder}$ (m)	$\delta_{G,\,max}$ (m)
None	2.249	−2.27	1.028/−1.028	12.681
161 GPa	1.716	−1.796	0.023/−0.036	12.675
168 GPa	1.715	−1.795	0.021/−0.035	12.675
200 GPa	1.712	−1.792	0.016/−0.030	12.674
206 GPa (Steel)	1.711	−1.791	0.016/−0.029	12.674
227 GPa	1.709	−1.789	0.013/−0.027	12.673
264 GPa	1.707	−1.787	0.01/−0.024	12.672
305 GPa	1.705	−1.786	0.007/−0.021	12.671
378 GPa	1.703	−1.784	0.004/−0.017	12.669
412 GPa	1.702	−1.783	0.003/−0.016	12.668

After the introduction of CFRP central buckles, the structural response is significantly improved. For instance, using CFRP with an elastic modulus of 161 GPa, the maximum end displacement of the girder is reduced to 1.716 m, representing a 23.7% decrease compared to the no-buckle case. This clearly demonstrates the effectiveness of the central buckle in limiting longitudinal displacement. Further analysis reveals a nonlinear relationship between displacement suppression efficiency and material stiffness. When the elastic modulus increases from 161 GPa to 412 GPa (a 156% increase), the maximum displacement is only slightly reduced from 1.716 m to 1.702 m—a reduction of merely 0.8%. This indicates that there exists a stiffness optimization threshold for the central buckle system, beyond which further increases in material stiffness yield diminishing returns in structural control effectiveness.

In addition, the CFRP cable with an elastic modulus of 161 GPa exhibits highly consistent displacement control performance compared to traditional steel cables, with a maximum displacement difference of only 0.005 m. This confirms the feasibility of using CFRP as a stiffness-equivalent substitute for steel. Given that CFRP has only 20% of the density of steel, its lightweight nature offers significant advantages in reducing structural self-weight, simplifying construction, and enhancing seismic performance.

The relative displacement between the cable and girder at midspan shows an even more pronounced trend than the girder-end displacement. As shown in Figures 5 and 10, in the absence of a central buckle, the relative displacement between the cable and girder reaches ±1.028 m. After introducing a CFRP central buckle with an elastic modulus of 161 GPa, this value significantly drops to +0.023 m and −0.036 m, representing an overall reduction of 97.8%. This result indicates that even low-stiffness FRP materials can effectively coordinate and constrain the deformation between cable and girder, demonstrating the high efficiency of the central buckle in suppressing relative displacement. Consistent with the conclusions drawn from longitudinal girder displacement control, the relative displacement suppression continues to improve with increasing CFRP stiffness but the rate of improvement diminishes progressively. This nonlinear relationship provides a critical basis for optimizing the material selection and stiffness configuration of the central buckle from both engineering and economic perspectives, avoiding resource waste and potential negative effects associated with over-stiffening.

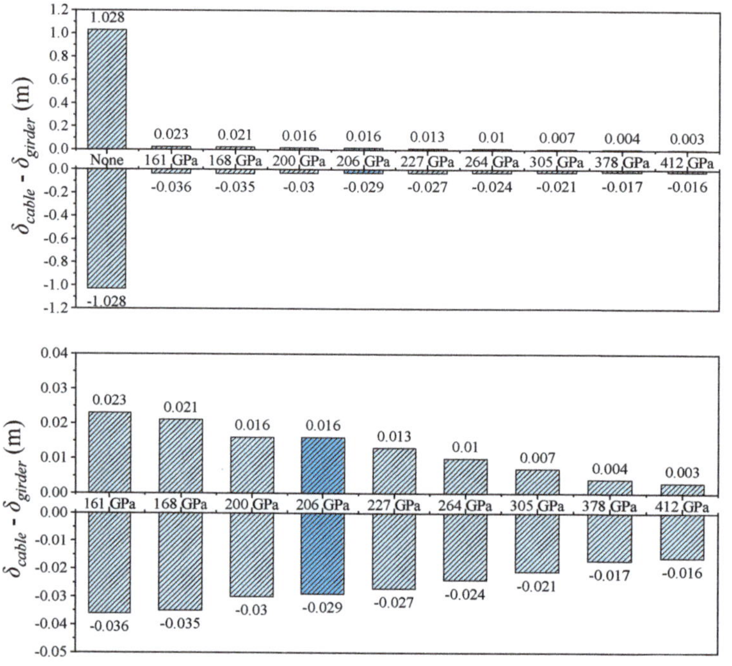

Figure 10. Midspan longitudinal relative displacement of cable and girder in static conditions.

Moreover, under all conditions, the maximum transverse displacement at the midspan of the main girder, $\delta_{G,\,max}$, remains stable, fluctuating within only 0.013 m (ranging from 12.668 m to 12.681 m). This indicates that variations in central buckle stiffness have negligible influence on transverse displacement, which is primarily governed by the torsional stiffness of the girder itself. Therefore, the design of the central buckle should focus on longitudinal control performance, while transverse displacement is largely dependent on the girder's overall torsional capacity.

4.1.2. Internal Forces of Central Buckles

Table 3 and Figure 11 present the axial force distribution of the three side members of the central buckle under the 161 GPa elastic modulus condition. The internal forces in B1, B2, and B3 are 7674 kN, 9810 kN, and 7100 kN, respectively. As the FRP material's elastic modulus increases from 161 GPa to 412 GPa, the axial forces in all three members exhibit nonlinear growth but the magnitude of increase varies significantly, revealing the spatial non-uniformity of internal force redistribution. Specifically, the force in B1 increases from 7674 kN to 8350 kN (an 8.8% increase), while B2 sees a substantial rise from 9810 kN to 14,323 kN (a 46.0% increase), and B3 experiences only a marginal increase to 7163 kN (a 0.9% increase). Among the three, B2 shows the highest sensitivity to changes in elastic modulus, indicating it serves as the primary load-bearing path. In contrast, the internal force responses of B1 and B3 are more influenced by local deformation compatibility mechanisms, making them less sensitive to stiffness variations.

Table 3. Internal forces of central buckles under varying stiffness levels.

E	B1 (kN)	B2 (kN)	B3 (kN)
None	/	/	/
161 GPa	7674	9810	7100
168 GPa	7703	9973	7105
200 GPa	7824	10,671	7119
206 GPa (Steel)	7844	10,790	7121
227 GPa	7913	11,286	7125
264 GPa	8022	12,103	7126
305 GPa	8128	12,928	7123
378 GPa	8286	14,104	7112
412 GPa	8350	14,559	7105

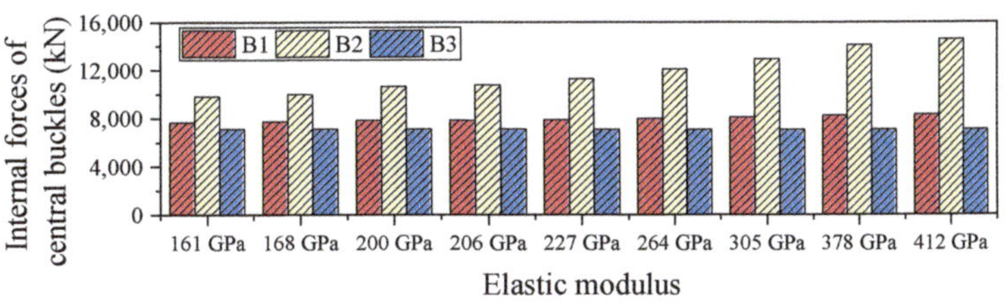

Figure 11. Internal forces of central buckles under varying stiffness levels in static conditions.

Overall, the maximum internal force increase in the central buckle (B2) reaches 46%, demonstrating the significant amplifying effect of material stiffness on internal force levels. However, in terms of displacement control, the improvement achieved by the high-modulus CFRP (412 GPa) over the low-modulus CFRP (161 GPa) is less than 1.2%. This indicates that the sensitivity of internal forces to stiffness enhancement is much greater than the marginal benefit in displacement control. Unless special performance requirements are present, excessively increasing the stiffness of the central buckle may significantly raise internal force levels, potentially leading to stricter demands on fatigue resistance and joint detailing. Considering both control effectiveness and structural safety, it is recommended to adopt CFRP cables with a relatively lower elastic modulus as the design choice for central buckle components.

4.1.3. Stress Amplitudes in Central Buckles and Hangers

Table 4 summarizes the stress amplitude responses of the central buckle and hangers under vehicular live load conditions for various elastic moduli. In suspension bridges, a typical negative correlation is observed between the stiffness of the flexible central buckle and the stress amplitude of hangers: as the stiffness of the central buckle increases, its ability to share live load improves, leading to a significant increase in its own stress amplitude while simultaneously helping to reduce the stress amplitude in adjacent hangers. Conversely, a central buckle with lower stiffness—though more deformable and experiencing smaller stress amplitudes itself—tends to transfer more load to the hangers, thereby increasing their stress amplitudes.

Table 4. Stress amplitudes in central buckles and hangers under vehicular live load.

E	B1 (MPa)	B1 (MPa)	B1 (MPa)	H1 (MPa)	H2 (MPa)	H3 (MPa)
None	/	/	/	115	115	115
161 GPa	87	166	96	90	89	123
168 GPa	88	175	97	93	88	123
200 GPa	93	198	101	108	85	126
206 GPa (Steel)	93	202	102	110	84	126
227 GPa	95	216	104	119	81	128
264 GPa	99	238	108	134	77	130
305 GPa	102	260	112	148	73	133
378 GPa	106	294	116	171	67	136
412 GPa	107	308	119	179	64	138

As shown in Figure 12, in the absence of a central buckle, the stress amplitude in all monitored hangers is 115 MPa, indicating a relatively uniform stress distribution when longitudinal restraint is lacking. After introducing a CFRP central buckle with an elastic modulus of 161 GPa, the stress distribution becomes markedly redistributed: the stress amplitudes at monitoring points 1 and 2 drop to 90 MPa and 89 MPa, respectively, representing reductions of 21.7% and 22.6%; however, the stress at point 3 rises to 123 MPa, an increase of 7.0%, indicating a differentiated influence of the central buckle on the force distribution among different hangers. When the CFRP elastic modulus is further increased to 412 GPa, the stress at point 1 significantly increases to 179 MPa, while that at point 2 decreases to 64 MPa, reflecting a more pronounced stress concentration effect resulting from the increased stiffness.

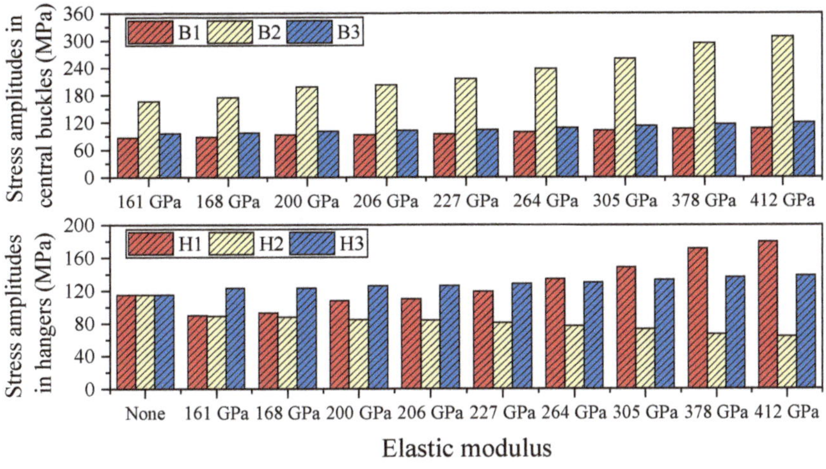

Figure 12. Stress amplitudes in central buckles and hangers under different stiffness levels in static conditions.

4.2. Seismic Response Analysis

4.2.1. Dynamic Properties

Figure 13 shows the first-order modal shapes of the suspension bridge without central buckles. As shown in Table 5, the fundamental frequency remains essentially constant at 0.0366 Hz as the central buckle stiffness increases, indicating that the central buckle has minimal influence on the lateral dynamic properties of the structure. This observation is consistent with the results reported by other researchers [15,29]. In contrast, the frequencies of the lateral and vertical bending modes show slight increases, with a maximum growth of approximately 3.5%, indicating relatively minor sensitivity to stiffness variation. Notably, the frequency of the first antisymmetric torsional mode increases significantly from 0.3084 Hz to 0.3461 Hz—an increase of 12.2%—as the central buckle stiffness increases. This indicates that this torsional mode is more sensitive to stiffness variation and that the structural torsional rigidity improves significantly with increased central buckle stiffness.

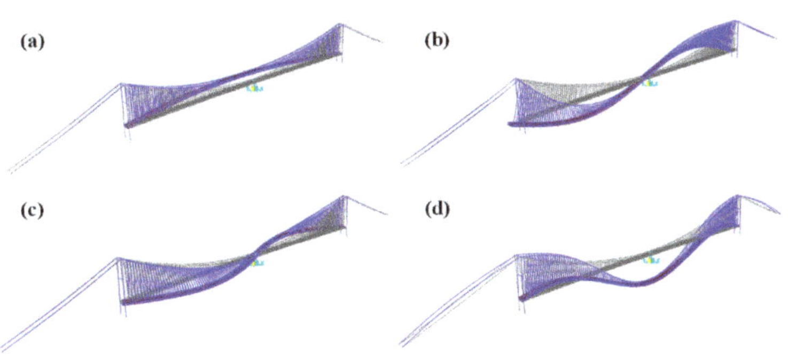

Figure 13. First-order modal shapes of the suspension bridge without central buckles: (**a**) symmetric lateral bending mode; (**b**) antisymmetric vertical bending mode; (**c**) antisymmetric lateral bending mode; (**d**) symmetric vertical bending mode.

Therefore, it can be concluded that the frequency responses of lateral and vertical primary modes are more strongly governed by structural geometry, while torsional modes are more sensitive to changes in central buckle stiffness.

Table 5. Fundamental frequencies under various central buckle stiffnesses.

E	1st Sym. Lat. Bend.	1st Antisym. Lat. Bend.	1st Sym. Vert. Bend.	1st Antisym. Vert. Bend.	1st Sym. Torsion	1st Antisym. Torsion
None	0.0366	0.0803	0.1015	0.0696	0.2291	0.3084
161 GPa	0.0366	0.0824	0.1016	0.0736	0.2291	0.3430
168 GPa	0.0366	0.0829	0.1016	0.0736	0.2291	0.3432
200 GPa	0.0366	0.0829	0.1017	0.0737	0.2291	0.3440
206 GPa (Steel)	0.03663	0.08292	0.10164	0.0737	0.2291	0.3439
227 GPa	0.0366	0.0830	0.1017	0.0737	0.2291	0.3444
264 GPa	0.0366	0.0830	0.1017	0.0737	0.2291	0.3449
305 GPa	0.0366	0.0830	0.1017	0.0737	0.2291	0.3453
378 GPa	0.0366	0.0831	0.1017	0.0737	0.2291	0.3459
412 GPa	0.0366	0.0831	0.1017	0.0737	0.2291	0.3461

4.2.2. Girder-End Displacement

Under seismic excitation, the longitudinal displacement response of the structure exhibits a distribution pattern similar to that under static loading. However, the peak displacements at the girder ends are significantly reduced, indicating the regulating effect of seismic inertial forces on structural deformation. As shown in Table 6 and Figures 14–16, in the absence of a central buckle, the maximum and minimum longitudinal displacements at the girder ends are 0.981 m and –1.080 m, respectively, and the relative displacement between the cable and girder at midspan reaches 0.883 m.

Table 6. Seismic displacement responses of the structure.

E	δ_{max} (m)	δ_{min} (m)	$\delta_{\text{cable}} - \delta_{girder}$ (m)	Base Moment at Tower (kN·m)
None	0.981	−1.080	0.883/−0.837	6,236,087
161 GPa	0.684	−0.749	0.133/−0.129	4,499,717
168 GPa	0.686	−0.751	0.123/−0.122	4,534,774
200 GPa	0.683	−0.784	0.102/−0.109	4,587,448
206 GPa (Steel)	0.681	−0.786	0.098/−0.108	4,594,717
227 GPa	0.674	−0.787	0.089/−0.102	4,648,096
264 GPa	0.661	−0.775	0.079/−0.091	4,692,674
305 GPa	0.645	−0.753	0.072/−0.079	4,704,726
378 GPa	0.623	−0.714	0.062/−0.063	4,635,268
412 GPa	0.615	−0.701	0.057/−0.058	4,577,461

After introducing a CFRP central buckle with an elastic modulus of 161 GPa, the peak displacement at the girder end decreases to 0.684 m (a reduction of 30.3%), while the minimum displacement improves to –0.749 m (a reduction of 30.7%). The relative displacement at midspan is significantly reduced to 0.134 m, representing a decrease of 84.9%. When the CFRP elastic modulus is increased to 412 GPa, the peak displacements at the girder ends further decrease to 0.615 m and –0.701 m, respectively. Compared to the 161 GPa case, this represents reductions of 10.1% and 6.4%, respectively, indicating that increased stiffness continues to suppress earthquake-induced displacements.

Consistent with the static loading results, the CFRP central buckle with a modulus of 161 GPa performs nearly identically to the conventional steel cable in terms of displacement control, with a maximum displacement difference of only 0.003 m. This further confirms the effectiveness of the FRP central buckle in practical engineering applications.

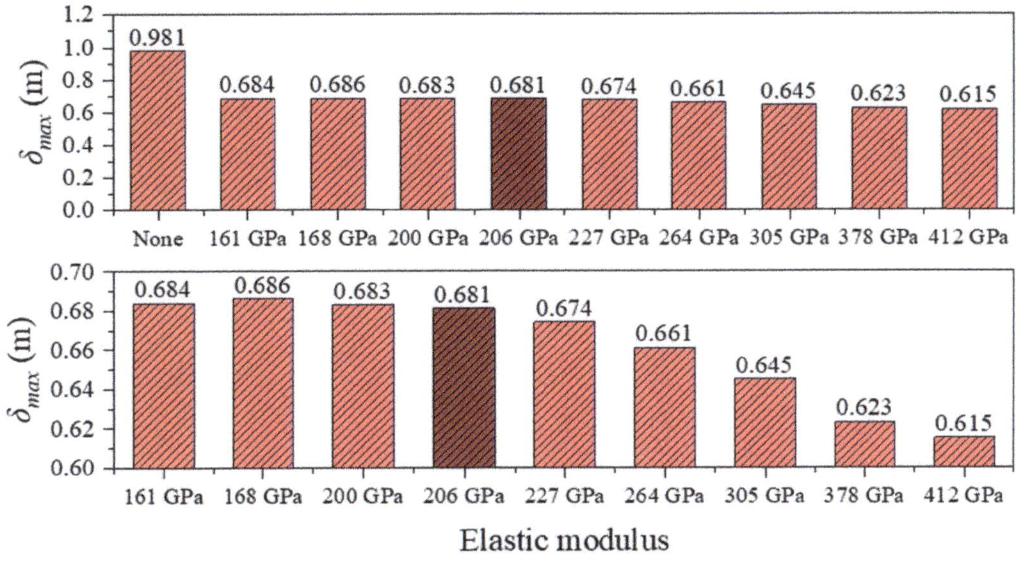

Figure 14. Maximum girder-end displacement under different central buckle stiffnesses in seismic conditions.

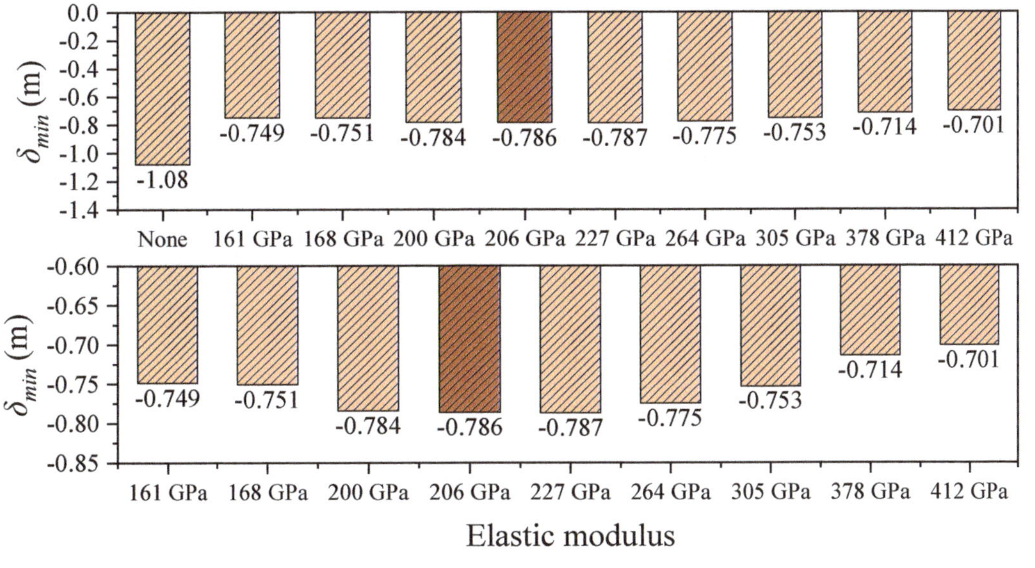

Figure 15. Minimum girder-end displacement under different central buckle stiffnesses in seismic conditions.

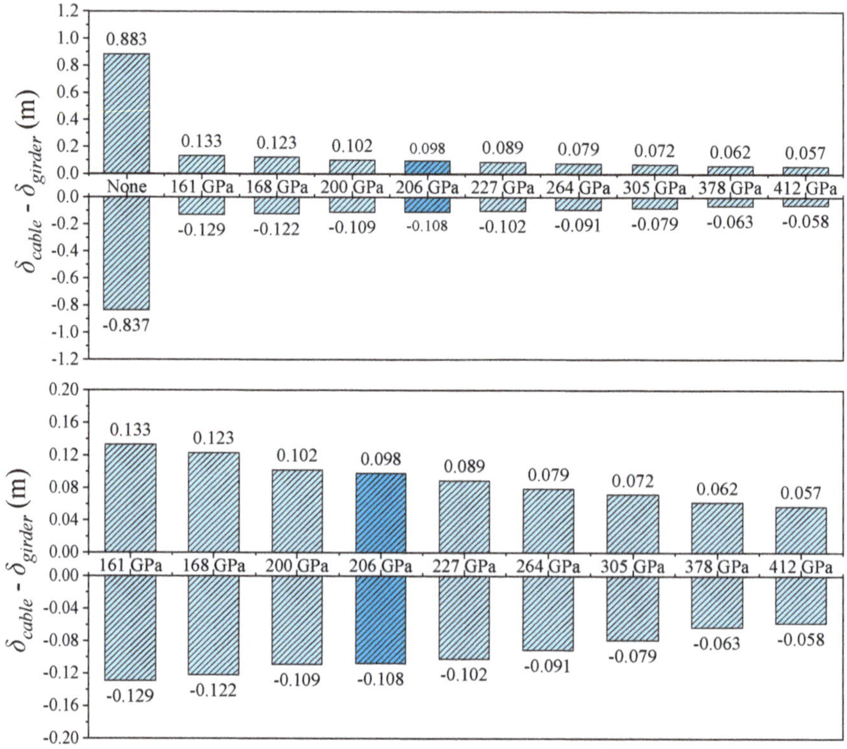

Figure 16. Midspan cable girder longitudinal relative displacement in seismic conditions.

4.2.3. Internal Force Response of Central Buckles

Under seismic loading, the internal force response of the FRP central buckle exhibits a distinct pattern compared to that under static loading conditions. Notably, with increasing stiffness, the internal force in B2 shows a decreasing trend. As shown in Table 7 and Figure 17, when the elastic modulus increases from 161 GPa to 412 GPa, the axial force in B2 decreases from 27,711 kN to 23,410 kN, representing a 15.5% reduction. This phenomenon is primarily attributed to the difference in loading characteristics. Under static loading, increased stiffness optimizes the load transfer path and enhances the internal force contribution of the central buckle. In contrast, under seismic excitation, although higher stiffness improves the mechanical involvement of the central buckle, the accompanying reduction in inertial mass—due to the fact that FRP has only about 20% of the density of steel—leads to a significant drop in member inertia. As a result, a "reverse evolution" trend emerges between stiffness increase and internal force response.

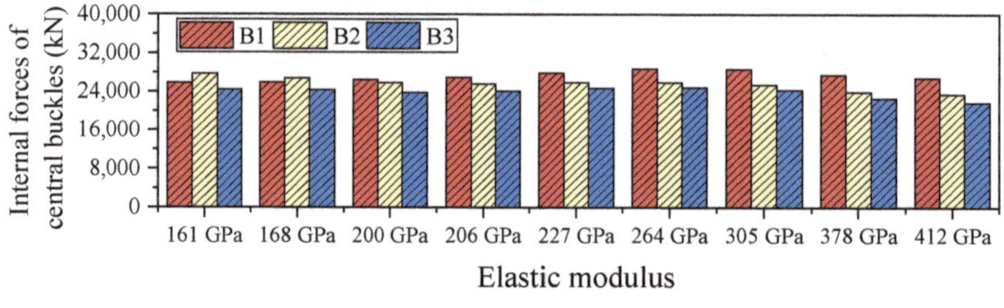

Figure 17. Internal forces of central buckles under varying stiffness levels in seismic conditions.

Table 7. Seismic internal force responses of central buckles.

E	B1 (kN)	B2 (kN)	B3 (kN)
None	/	/	/
161 GPa	25,871	27,711	24,428
168 GPa	25,889	26,799	24,386
200 GPa	26,498	25,875	23,820
206 GPa (Steel)	26,920	25,604	24,135
227 GPa	27,847	25,909	24,711
264 GPa	28,722	25,917	24,900
305 GPa	28,646	25,443	24,342
378 GPa	27,507	23,952	22,551
412 GPa	26,833	23,410	21,641

The evolution of the tower base bending moment reflects the regulatory effect of central buckle stiffness on the overall load transfer system of the suspension bridge. In the absence of a central buckle, the longitudinal bending moment at the tower base reaches as high as 6.24×10^6 kN·m (Figure 18). After introducing a CFRP central buckle with an elastic modulus of 161 GPa, the tower base moment is significantly reduced by 27.8%, decreasing to 4.50×10^6 kN·m. The core mechanism lies in the central buckle's ability to coordinate deformation between the main cable and the girder, thereby reconstructing the load transfer path. The presence of the central buckle causes a greater portion of the cable tension to be concentrated in the midspan region of the bridge, effectively reducing the bending moment that must be resisted by the main towers.

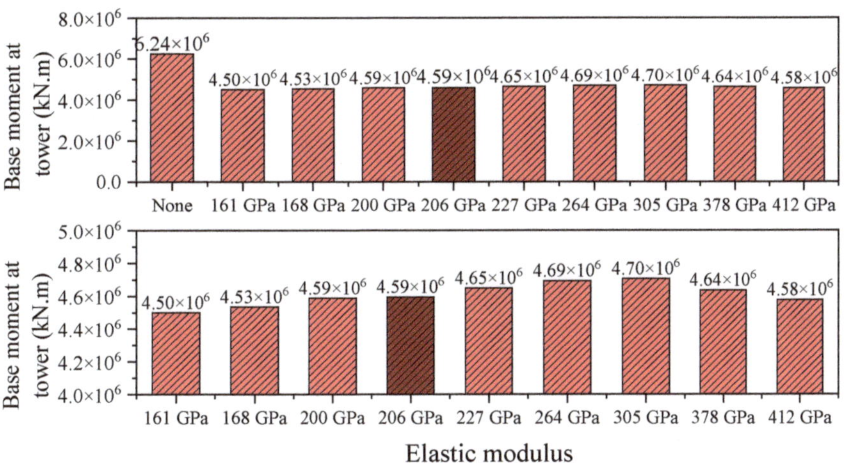

Figure 18. Base moment at tower under different central buckle stiffnesses in seismic conditions.

4.3. Cost Analysis

In bridge engineering practice, cost-effectiveness is a key factor influencing the material selection and structural configuration of the central buckle. Although Carbon Fiber-Reinforced Polymer (CFRP) central buckles offer significant advantages over traditional steel systems in terms of mechanical performance and durability, their material costs can vary widely depending on the type and grade of carbon fiber used. To balance performance requirements with economic feasibility, this study conducts a comparative analysis of the material costs associated with different types of CFRP central buckles. For simplification, only the unit price of raw carbon fiber materials is considered, excluding additional costs such as fabrication, transportation, and installation. Table 8 presents the reference market prices of various types of carbon fibers, based on engineering procurement data in China, updated to the year 2025.

Table 8. Unit cost comparison of different CFRP strands.

Type	T300	T700G	T800S	T1000S	M40J	M46J	M55J	M60J
Cost (USD/kg)	30	50	60	1000	75	80	120	150

To assess the practical applicability of FRP central buckles with different stiffness levels, a systematic comparison was performed under both static and seismic loading conditions, considering material cost, internal force, and maximum longitudinal displacement at the girder ends. As illustrated in Figure 19, although high-modulus CFRP central buckles can enhance the longitudinal stiffness of suspension bridges and suppress displacement responses to some extent, the improvement in displacement control is minimal. Moreover, they induce significantly higher internal forces and lead to increased material costs. In contrast, low-modulus CFRP central buckles with an elastic modulus of 161 GPa demonstrate an excellent performance–cost balance under both static and seismic conditions. Their displacement control effectiveness is comparable to that of medium- and high-modulus materials (with differences less than 1 mm), while generating significantly lower internal forces, thereby offering greater structural safety margins. Furthermore, this type of CFRP features the lowest unit cost, effectively reducing total material expenses. Especially under seismic conditions, low-modulus central buckles help mitigate inertial force amplification caused by excessive stiffness, further optimizing the force distribution in towers and cables.

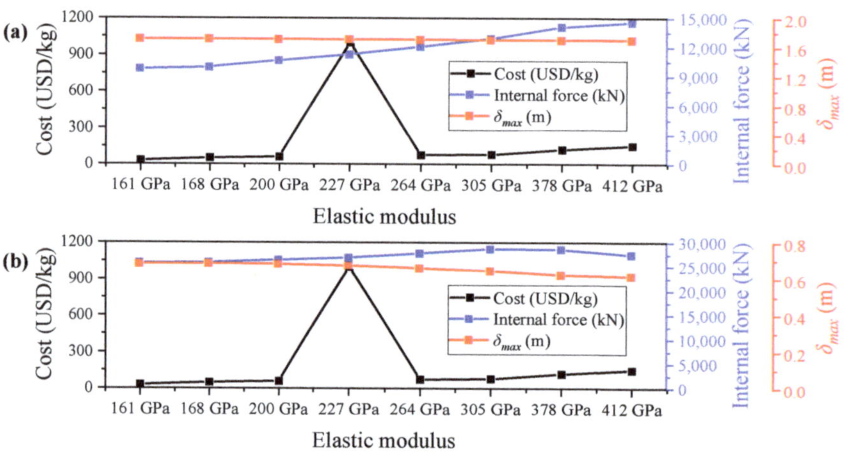

Figure 19. Performance–cost trade-off analysis of CFRP central buckles: (**a**) static load case; (**b**) seismic load case.

In summary, low-modulus CFRP materials (such as T300) not only provide sufficient deformation control capabilities but also offer significant advantages in internal force reduction and economic efficiency. They represent a highly cost-effective option for central buckle design in ultra-long-span suspension bridge projects.

In addition to the comparison of material unit prices, it is important to note the potential advantages of CFRP central buckles over traditional steel counterparts in terms of life-cycle performance. Although the initial cost of CFRP materials may be higher, their inherent corrosion resistance can significantly reduce long-term maintenance demands, especially in marine or salt-laden environments. Furthermore, the lighter weight of CFRP components facilitates transportation and installation, potentially lowering labor and equipment costs. The superior fatigue-resistance of CFRP also contributes to an extended service life with reduced structural deterioration under repeated loading.

4.4. Combined Effect of CFRP Central Buckles and Girder-End Restraints

Although the central buckle can effectively reduce the relative displacement between the cable and girder in suspension bridges, the longitudinal displacement at the girder ends can still reach ±1.7 m under static loading, which exceeds the design threshold of conventional expansion joints. To address this issue, this study proposes a collaborative control strategy that combines CFRP central buckles with end-girder restraining devices, aiming to overcome the performance limitations of a single component. A parametric combination analysis was conducted to systematically investigate the synergy between central buckles and restraining devices of varying stiffness. To clarify the mechanical response differences among different configurations, four representative scenarios were defined: (1) Scenario S1: three pairs of 161 GPa CFRP central buckles without restraining devices; (2) Scenario S2: three pairs of 200 GPa CFRP central buckles without restraining devices; (3) Scenario S3: three pairs of 161 GPa CFRP central buckles with restraining devices; (4) Scenario S4: three pairs of 200 GPa CFRP central buckles with restraining devices. The response characteristics of each configuration are detailed in Table 9.

Table 9. Results of combined application.

Load Condition	Items	S1	S2	S3	S4
Static	Girder-end displacement (m)	1.795	1.791	1.103	1.103
	Cable–girder relative displacement (m)	0.035	0.029	0.035	0.029
	Internal forces of central buckles (kN)	9973	10,790	9973	10,790
	Stress amplitudes in central buckles (MPa)	175	202	175	202
	Stress amplitude of hangers (MPa)	123	126	123	126
	Restraint force (kN)	/	/	34,738	34,994
Seismic	Girder-end displacement (m)	0.751	0.786	0.292	0.294
	Cable–girder relative displacement (m)	0.123	0.108	0.062	0.053
	Moment at tower base (kN·m)	4,534,774	4,594,717	4,150,151	4,177,035
	Damping force (kN)	/	/	2848	2853
	Internal forces of central buckles (kN)	26,799	26,920	13,450	13,902

Figure 20 shows that the restraining devices have a significant effect on controlling the longitudinal displacement of the main girder in suspension bridges. Under static loading, when using 161 GPa CFRP central buckles, the introduction of restraining devices reduces the longitudinal displacement at the girder ends from 1.790 m to 1.103 m—a reduction of 38.4%. Under seismic loading, the control effect is even more pronounced, achieving a 61.1% reduction, while the midspan cable–girder relative displacement is also reduced by 49.6%. These results demonstrate the strong suppression capability of restraining devices under dynamic loads. The core mechanism behind this displacement control lies in the restraining devices enhancing the longitudinal stiffness of the girder, thereby reconstructing the path through which inertial forces are transmitted. Part of the longitudinal inertial force is diverted directly to the abutment foundations instead of being entirely transmitted through the central buckle. Notably, this load path redistribution not only effectively suppresses large longitudinal displacements of the girder but also significantly alleviates the internal forces in the central buckle and towers. For example, under seismic loading, when using 161 GPa CFRP central buckles, the introduction of restraining devices reduces the axial force in Member B2 of the central buckle from 26,799 kN to 13,450 kN—a reduction of 49.8%. Although the restraining blocks effectively limit longitudinal displacement, they are subjected to significant reaction forces. Under the static S3 loading condition, the peak reaction force on the restraining block reaches 34,738 kN. This introduces challenges in the detailed design of the restraining components.

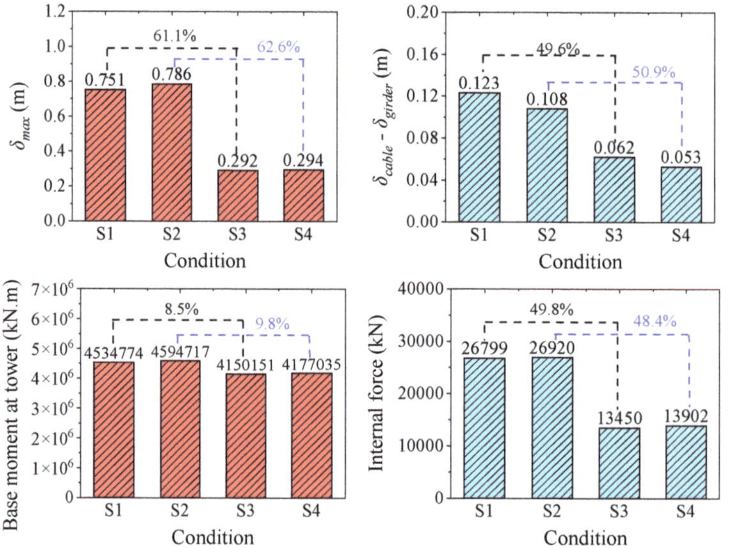

Figure 20. Seismic response results of the combined application scheme.

In summary, the combination of low-modulus CFRP central buckles and end-girder restraining devices demonstrates superior collaborative control performance and represents one of the most effective solutions for managing longitudinal displacement and internal forces in suspension bridges.

In addition to reducing instantaneous displacement and internal forces, the combined use of CFRP central buckles and end-span restraining devices may also benefit the long-term behavior of the bridge. By increasing longitudinal stiffness at the girder ends, the system modifies internal load paths and reduces sustained stress on the central buckle and nearby hangers. This leads to lower stress amplitudes and frequencies, which may help delay fatigue damage under repeated traffic, temperature variation, and seismic loading. While fatigue analysis is beyond the scope of the current study, the findings provide a basis for future studies on long-term performance under cumulative effects.

5. Conclusions

In this paper, a novel CFRP central buckle system for ultra-long-span steel truss suspension bridges is proposed. A comprehensive finite element analysis was conducted to investigate the influence of central buckle stiffness on structural performance, and a performance–cost evaluation framework was developed to assess its engineering applicability. The main conclusions are as follows:

1. The CFRP central buckle with an elastic modulus of 161 GPa provides displacement control performance comparable to that of traditional steel central buckles, while significantly mitigating stress concentration issues. With additional advantages such as low weight, high strength, and excellent durability, CFRP central buckles show strong application potential in long-span bridge engineering.

2. CFRP central buckles exhibit a typical nonlinear mechanical response. Although increasing stiffness can enhance structural control capacity, a threshold effect is observed. When the elastic modulus increases from 161 GPa to 412 GPa, the displacement of the main girder is reduced by only 0.8%, indicating diminishing returns. This highlights the need to seek an optimal balance between performance and cost.
3. Comprehensive analysis demonstrates that low-stiffness CFRP central buckles achieve a favorable balance among displacement control, load redistribution, and material cost reduction. These findings provide both theoretical and design references for the broader application of high-performance composites in critical bridge components.
4. A collaborative control strategy combining the CFRP central buckle with end-girder restraining devices is proposed and validated. When using a 161 GPa CFRP buckle, the combined system reduces maximum girder-end displacement by up to 61.1% while also decreasing internal forces in the central buckle and tower base bending moments. This results in a more efficient and integrated configuration strategy.

Future work may incorporate experimental studies and full-scale bridge monitoring to further evaluate the global structural performance and long-term service behavior of suspension bridges equipped with CFRP central buckles under complex loading environments. Although this study focuses on a long-span suspension bridge, the underlying control mechanism of the CFRP central buckle is applicable to other suspension bridge configurations with varying spans, girder types, and loading conditions. However, the optimal stiffness and layout must be adjusted based on the specific structural and dynamic characteristics. Further studies are recommended to evaluate performance under diverse boundary conditions and support broader engineering applications.

Author Contributions: Conceptualization, M.W. and Y.L.; methodology, Y.L.; software, H.Y.; validation, M.W., Y.L. and Y.Z.; formal analysis, H.Y.; investigation, H.Y.; resources, T.Z.; data curation, H.Y.; writing—original draft preparation, B.L.; writing—review and editing, Y.L. and M.W.; visualization, M.W.; supervision, Y.Z.; project administration, M.W.; funding acquisition, M.W. and T.Z. All authors have read and agreed to the published version of the manuscript.

Funding: This research was funded by the National Key Research and Development Program of China, grant number 2021YFB3704400; by the National Key Research and Development Program of China, grant number 2022YFC3801800; and the Science and Technology Project of Guangdong Provincial Communications Group Co., Ltd., grant number JT2023ZD01-01.

Institutional Review Board Statement: Not applicable.

Data Availability Statement: The original contributions presented in this study are included in the article. Further inquiries can be directed to the corresponding author.

Conflicts of Interest: Author Maoqiang Wang, Huaimao Yang, Yaoyu Zhu were employed by the company CCCC Highway Bridges National Engineering Research Centre, Ltd. Author Taike Zhang was employed by the company Guangdong Provincial Highway Construction Co., Ltd. and Guangdong Bay Area Transportation Construction Investment Co., Ltd. The remaining authors declare that the research was conducted in the absence of any commercial or financial relationships that could be construed as a potential conflict of interest. The funder was consulted regarding the study design and material selection but had no role in the collection, analyses, interpretation of data, writing of the manuscript and the decision to publish the results.

References

1. Guo, T.; Liu, J.; Huang, L. Investigation and control of excessive cumulative girder movements of long-span steel suspension bridges. *Eng. Struct.* **2016**, *125*, 217–226. [CrossRef]
2. Wang, Z.; Peng, Y.; Lan, X.; Bai, X.; Deng, C.; Ren, Y. Selection and parameter optimization of constraint systems for girder-end longitudinal displacement control in three-tower suspension bridges. *Struct. Durab. Health Monit.* **2025**, *19*, 643–664. [CrossRef]

3. Hu, J.; Wang, L.; Song, X.; Sun, Z.; Cui, J.; Huang, G. Field monitoring and response characteristics of longitudinal movements of expansion joints in long-span suspension bridges. *Measurement* **2020**, *162*, 107933. [CrossRef]
4. Zhang, W.-M.; Chen, Y.-P. An analytical method for assessing the live load deformation of a suspension bridge with an improved central buckle. *Structures* **2023**, *47*, 1341–1351. [CrossRef]
5. Cao, H.; Chen, Y.; Li, J.; Liu, S. Static characteristics analysis of three-tower suspension bridges with central buckle using a simplified model. *Eng. Struct.* **2021**, *245*, 112916. [CrossRef]
6. Liu, W.; Song, S.K.; Quan, X.R.; Peng, Y.S.; Qi, Q.M. Seismic performance study of BRB central buckles in suspension bridges. *Bridge Constr.* **2022**, *52*, 61–68. (In Chinese)
7. Xu, X.; Qiang, S.Z. Study on the influence of central buckles on dynamic characteristics and seismic response of long-span suspension bridges. *J. China Railw. Soc.* **2010**, *32*, 84–91. (In Chinese)
8. Zhang, W.; Tian, G.; Chen, Y. Evolution of suspension bridge structural systems, design theories, and shape-finding methods: A literature survey. *J. Traffic Transp. Eng. (Engl. Ed.)* **2024**, *11*, 225–244. [CrossRef]
9. Chen, Z.H.; Chen, L.; Shao, C.Y. Study on the mechanical performance of multiple rigid central buckles in long-span suspension bridges. *Eng. Mech.* **2024**, *41*, 19–28. (In Chinese) [CrossRef]
10. Guo, Z.M.; Wang, H.X.; Ye, A.J. Study on longitudinal seismic resistance system of extra-long-span suspension bridges with flexible central buckles. *Bridge Constr.* **2020**, *50*, 38–43. (In Chinese)
11. Qin, F.J.; Di, J.; Dai, J.; Li, G.L. Influence of central buckle on dynamic behavior of long-span suspension bridge with deck-truss composite stiffening girder. *Adv. Mater. Res.* **2013**, *838–841*, 1096–1101. [CrossRef]
12. Xiao, H.Z.; Xie, L.B.; Qiu, F. Study on end displacement characteristics of main girders in long-span cable-supported bridges. *Bridge Constr.* **2023**, *53* (Suppl. S2), 72–78. (In Chinese)
13. Xiao, Y.G.; Chen, W.H. Parametric analysis of dynamic characteristics of long-span suspension bridges. *J. Chang. Univ. Sci. Technol. (Nat. Sci. Ed.)* **2016**, *13*, 28–33. (In Chinese)
14. Zhao, Y.; Guo, X.; Su, B.; Sun, Y.; Li, X. Evaluation of flexible central buckles on short suspenders' corrosion fatigue degradation on a suspension bridge under traffic load. *Materials* **2023**, *16*, 290. [CrossRef]
15. Wang, D.; Deng, Y.; Liu, Y. Influence of central buckle on suspension bridge dynamic characteristics and driving comfort. *J. Cent. South Univ.* **2015**, *22*, 3108–3115. [CrossRef]
16. Song, H.; Lin, Q.; Tang, M.L. Analysis of the effect of central buckles on suspension bridges with different cable systems. *Eng. Sci.* **2010**, *12*, 22–27. (In Chinese)
17. Zhou, G.L.; Li, X.J.; Liu, B.D. Application, error analysis and improvement of large mass method in multi-support excitation analysis. *Eng. Mech.* **2011**, *28*, 48–54. (In Chinese)
18. Qin, Q.; Lou, L. Influence of non-classical damping on seismic response of suspension bridges. *China Civ. Eng. J.* **2000**, *3*, 17–22. (In Chinese) [CrossRef]
19. Zhang, W.-M.; Shen, X.-H.; Chang, J.-Q. Main cable shape-finding and live load response of the suspension bridge with central buckles: An analytical algorithm. *J. Bridge Eng.* **2023**, *28*, 04023084. [CrossRef]
20. Guo, T.; Liu, J.; Zhang, Y.; Pan, S. Displacement monitoring and analysis of expansion joints of long-span steel bridges with viscous dampers. *J. Bridge Eng.* **2015**, *20*, 04014099. [CrossRef]
21. Liu, Y.; Zhang, H.-T.; Tafsirojjaman, T.; Ur Rahman Dogar, A.; AlAjarmeh, O.; Yue, Q.-R.; Manalo, A. A novel technique to improve the compressive strength and ductility of glass fiber reinforced polymer composite bars. *Constr. Build. Mater.* **2022**, *326*, 126782. [CrossRef]
22. Liu, Y.; Zhang, H.-T.; Tafsirojjaman, T.; Ur Rahman Dogar, A.; Yue, Q.-R.; Manalo, A. Compressive behaviour and prediction model for short and slender FRP-confined GFRP bars. *Constr. Build. Mater.* **2023**, *376*, 131059. [CrossRef]
23. Liu, Y.; Tafsirojjaman, T.; Dogar, A.U.R.; Hückler, A. Shrinkage behavior enhancement of infra-lightweight concrete through FRP grid reinforcement and development of their shrinkage prediction models. *Constr. Build. Mater.* **2020**, *258*, 119649. [CrossRef]
24. Liu, Y.; Zwingmann, B.; Schlaich, M. Carbon fiber reinforced polymer for cable structures—A review. *Polymers* **2015**, *7*, 2078–2099. [CrossRef]
25. Liu, Y.; Tafsirojjaman, T.; Dogar, A.U.R.; Hückler, A. Bond behaviour improvement between infra-lightweight and high strength concretes using FRP grid reinforcements and development of bond strength prediction models. *Constr. Build. Mater.* **2021**, *270*, 121426. [CrossRef]
26. He, W.; Zhou, Y.; Xu, H.; Ding, Z. Multi-objective optimization and shaking table test of adjacent structure with viscous dampers. *J. Bridge Eng.* **2025**, *100*, 111730. [CrossRef]
27. *JTG 3363-2019*; Ministry of Transport of the People's Republic of China. Specifications for Design of Highway Bridge Foundation. People's Transportation Publishing House: Beijing, China, 2019. (In Chinese)

28. *JTG B01-2014*; Ministry of Transport of the People's Republic of China. Technical Standard of highway Engineering. People's Transportation Publishing House: Beijing, China, 2014. (In Chinese)
29. Hu, T.-F.; Hua, X.-G.; Zhang, W.-W.; Xian, Q.-S. Influence of central buckles on the modal characteristics of long-span suspension bridge. *J. Highw. Transp. Res. Dev. (Engl. Ed.)* **2016**, *10*, 72–77. [CrossRef]

polymers

MDPI

Article

Study on the Influence and Mechanism of Steel, Polyvinyl Alcohol, and Polyethylene Fibers on Slag–Yellow River Sediment Geopolymers

Ge Zhang [1,2,3], Enhui Jiang [1,2,3], Kunpeng Li [1,2,3,*], Huawei Shi [1,2,3], Chen Chen [1,2,3] and Chengfang Yuan [4,*]

1 Yellow River Institute of Hydraulic Research, Yellow River Water Conservancy Commission, Zhengzhou 450003, China; gezhangyrihr@163.com (G.Z.); jiangenhui@hky.yrcc.gov.cn (E.J.); 15538352232@163.com (H.S.); 15617633649@163.com (C.C.)
2 Key Laboratory of Lower Yellow River Channel and Estuary Regulation, Ministry of Water Resources, Zhengzhou 450003, China
3 Yellow River Laboratory, Zhengzhou 450003, China
4 College of Civil Engineering, Zhengzhou University, Zhengzhou 450001, China
* Correspondence: likunpeng@hky.yrcc.gov.cn (K.L.); chengfang1102@zzu.edu.cn (C.Y.)

Abstract: Steel fibers (STs), polyvinyl alcohol fibers (PVAs), and polyethylene fibers (PEs) were selected to systematically investigate the effects of different fiber types and dosages on the workability (slump and spread) and mechanical properties (compressive strength and splitting tensile strength) of slag–Yellow River sand geopolymer eco-cementitious materials. By combining microstructural testing techniques such as thermogravimetric-differential thermal analysis (TG-DTA), X-ray diffraction (XRD), and scanning electron microscopy-energy dispersive spectroscopy (SEM-EDS), the influence mechanisms of fibers on the characteristic products and microstructure of the matrix were thoroughly revealed, and the role of fibers in the strength development of Yellow River sediment-based geopolymers was elucidated. The results show that as the fiber content increases, the workability of the mixture significantly decreases. The appropriate incorporation of steel fibers and PVAs can significantly enhance the strength and toughness of the matrix. When the fiber dosage is 1%, the 28-day compressive strength of specimens with steel fibers and PVAs increased by 25.93% and 21.96%, respectively, compared to the control group, while the splitting tensile strength increased by 50.00% and 60.34%, respectively. However, the mechanisms of action differ significantly; steel fibers primarily enhance the compressive performance of the matrix through their high stiffness and strength, whereas PVAs inhibit crack propagation through their flexibility and excellent bonding properties. In contrast, the strength improvement of PEs is mainly reflected in toughening. When the fiber dosage is 1.5%, the 28-day splitting tensile strength of PE specimens increased by 72.61%, and the tensile-to-compressive ratio increased by 92.32% compared to the control group. Microstructural analysis indicates that the incorporation of different types of fibers does not alter the types of characteristic products in alkali-activated cementitious materials, but excessive fiber content affects the generation of gel-like products and the distribution of free water, thereby altering the thermal decomposition behavior of characteristic gel products. Additionally, the matrix incorporating PEs forms a honeycomb-like amorphous gel, resulting in weak interfacial bonding between the fibers and the matrix. This is one of the main reasons for the limited reinforcing effect of PEs at the microscopic scale and a key factor for their inferior long-term performance compared to steel fibers and PVAs. This study provides theoretical foundations and practical guidance for optimizing the performance of fiber-reinforced geopolymer materials.

https://doi.org/10.3390/polym17081072

Keywords: Yellow River sediment (YRS); geopolymer; fiber; strength; microstructure

1. Introduction

The protection of the Yellow River is crucial for the long-term revitalization of China. The Yellow River basin faces significant sedimentation challenges due to factors such as low water volume, excessive sediment, and imbalanced water–sediment relationships. Sediment deposition presents a critical technical challenge for flood control, maintaining reservoir storage capacity, and ensuring the safe operation of irrigation systems in the region. However, the demand for construction sand in China is enormous, with approximately 20 billion tons of sand and gravel used annually [1]. Consequently, natural sand resources are increasingly depleted, necessitating the search for stable alternatives. Against this backdrop, the resource attributes and economic value of Yellow River sediment have become prominent. The main component of Yellow River sediment is silicon dioxide (SiO_2) and based on this component alone Yellow River sediment is a potential alternative material. Utilizing Yellow River sediment resources offers significant socio-economic and ecological environmental benefits.

Ordinary Portland cement (OPC) is the most commonly used cementitious material in the construction industry. Its extensive use has contributed to global infrastructure and economic development to a certain extent. However, the OPC production process causes significant harm to the natural environment [2]. This highlights the urgent need to develop low-carbon, high-performance cementitious materials for infrastructure needs. Geopolymers, also known as alkali-activated cementitious materials, are characterized by a rapid setting speed [3,4], high-temperature resistance [5–7], corrosion resistance [8–10], and acid resistance [11,12]. Geopolymers can be prepared without the need for calcination, offering great potential to reduce carbon emissions by replacing cement [13–16]. However, similar to conventional concrete, geopolymer materials suffer from high brittleness [17,18] and low tensile strength [19,20], which limits their widespread application [21,22].

Incorporating fibers into geopolymer materials can significantly enhance their engineering properties, such as their tensile strength, flexural strength, shrinkage behavior, and resistance to fatigue, impact, and thermal shock [23,24]. Nowadays, various fibers, including polypropylene fibers (PPs) [25–27], polyamide fibers (PAs), polyvinyl alcohol fibers (PVAs) [28–30], steel fibers (STs) [31–33], and glass (GL) fibers [34,35], have been widely used, each offering potential benefits or drawbacks to concrete properties. Lee et al. [36] investigated slag-based alkali-activated mortar reinforced geopolymers with oiled PVAs and demonstrated the feasibility of achieving a tensile strain of up to 4.7%. Natali et al. [37] modified some properties of alkali-activated ladle-slag at 7 d by employing PVAs, and the enhancement in ductility after the first crack load was obvious. Zhang et al. [38,39] investigated the behavior of the short PVA-reinforced fly ash-metakaolin geopolymer boards at 28 d. Their conclusion demonstrated that the incorporation of high-volume PVAs changed the impact failure mode of geopolymer boards from a brittle to ductile pattern. Feng Hu et al. [40]. utilized fly ash and PEs in the preparation of engineered geopolymer composites (EGCs), and the resulting material has been effectively implemented in the construction of subway tunnel shield segments. Yuan et al. [41]. investigated Yellow River sand (YRS) in engineered cementitious composites (ECCs), finding that 100% YRS substitution enhances strength, with the compressive strength and flexural strength increasing by 26.4 MPa and 11.5 MPa, respectively, while 75% substitution provides optimal ductility, promoting sustainable concrete applications. Akturk et al. [42] demonstrated that steel fiber or polypropylene fiber addition improved the flexural strength of sodium carbonate-activated slag mortars. Zhou et al. [43] showed that increasing basalt fiber dosage enhanced the flexural and splitting tensile strength of fiber-reinforced alkali-activated slag concrete. Abdollahnejad et al. [44] incorporated various fiber types into alkali-activated slag mortar, finding improvements in flexural strength, while the effect on compressive strength varied depending on the fiber type.

The Yellow River sediment significantly differs from conventional river sand used in construction materials in terms of particle gradation and physicochemical properties. The sediment particles are predominantly fine, falling within the silt and clay range (<0.075 mm), with shapes that are mostly flaky or angular, featuring rough surfaces and a large specific surface area. Additionally, the sediment has a high mud content (typically exceeding 10%). These characteristics result in notable differences in the performance of geopolymer materials prepared using Yellow River sediment as an aggregate compared to those using conventional river sand. Consequently, the effects of fibers on the performance of Yellow River sediment-based geopolymer ecological cementitious materials cannot be entirely extrapolated from existing studies. Targeted adaptive research and exploration are necessary to understand and optimize their performance. However, there is limited research on the effects of various fibers on the properties of Yellow River sediment geopolymers and the underlying mechanisms.

Based on the aforementioned background, this paper selects steel fibers (hereinafter referred to as STs), polyvinyl alcohol fibers (hereinafter referred to as PVAs), and polyethylene fibers (hereinafter referred to as PEs) to systematically investigate the influence of different fiber types and dosages on the workability and strength of slag–Yellow River sediment geopolymer ecological cementitious materials by setting varying dosages. Combined with micro-testing methods such as TG-DTA, XRD, and SEM-EDS, the influence mechanisms of STs, PVAs, and PEs on the characteristic products and microstructure of slag–Yellow River sediment geopolymer ecological cementitious materials will be explored. The study aims to provide valuable insights into the engineering applications of fiber-reinforced slag–Yellow River sediment geopolymer materials.

2. Materials and Methods

2.1. Raw Material and Mixed Proportion

2.1.1. Raw Material

The raw materials used in the experiment include Yellow River sediment (Abbreviate as YRS), slag, hybrid fiber, NaOH, and water glass. The Yellow River sediment was collected from the Xixiayuan Reservoir in Henan Province, initially in a moist state, and then dried before being used as a raw material. The chemical composition and particle size analysis of the YRS and slag were conducted using XRF and the Malvern Mastersizer-2000 laser particle size analyzer (UK). The results are shown in Table 1 and Figure 1, respectively. The slag used in this experiment is of grade S105, with a median particle size of 10.28 μm, a hydraulic coefficient of 2.17, an activity coefficient of 0.48, and an alkaline coefficient of 1.14. SEM images, shown in Figure 2, revealed that YRS particles displayed irregular geometric shapes, with considerable variation in particle size. Slag particles demonstrated a heterogeneous morphology, consisting of irregular shapes, flakes, spheres, and angular fragments. Three hybrid fibers were selected, namely short and fine copper-plated steel fibers, short-cut PVAs, and PEs, the specific performance indicators of which are shown in Table 2. Water glass, commonly known as sodium silicate, is a general term for $Na_2O{\cdot}nSiO_2$, where n is usually referred to as the modulus. It is a transparent glassy solution composed of alkali metal silicates. The chemical composition of water glass essentially depends on the molecular ratio n between SiO_2 and alkali metal oxides (Na_2O or K_2O), known as the water glass modulus. In this experiment, liquid sodium silicate produced by Shandong Yourui

Chemical Co., Ltd. (Zibo, China) was used, with the physicochemical parameters shown in Table 3. Pure NaOH was used to adjust the water glass molar ratio to 1.2, and the Na_2O content of the binder was set to 5%. Tap water was used for mixing in the experiment. The modulus adjustment equation is shown in Formula (1), which follows:

$$Na_2O{\cdot}2.18SiO_2 + NaOH \rightarrow Na_2O{\cdot}nSiO_2 + H_2O \quad (1)$$

Table 1. Chemical compositions of YRS and slag (wt.%).

Minerals	SiO_2	CaO	Al_2O_3	Fe_2O_3	K_2O	TiO_2	MgO	Other
YRS	68.64	8.40	12.33	3.25	2.55	0.74	2.05	2.04
slag	32.47	41.06	14.52	0.28	0.44	1.25	7.08	2.9

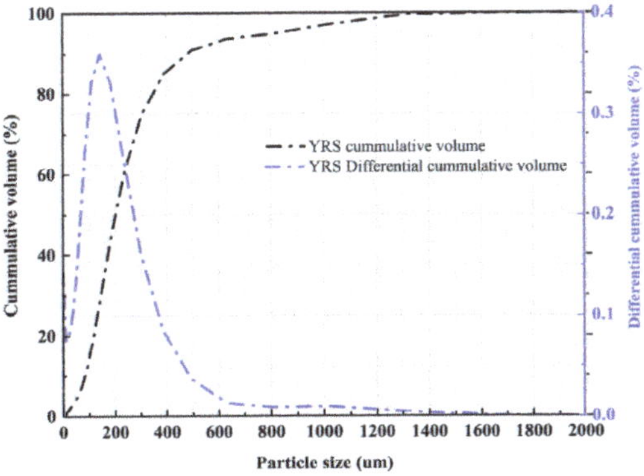

(a) YRS

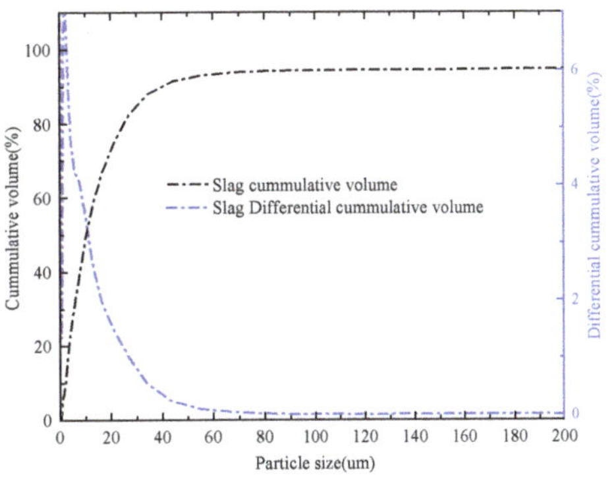

(b) Slag

Figure 1. Particle size distribution curves of YRS and slag.

(a) YRS

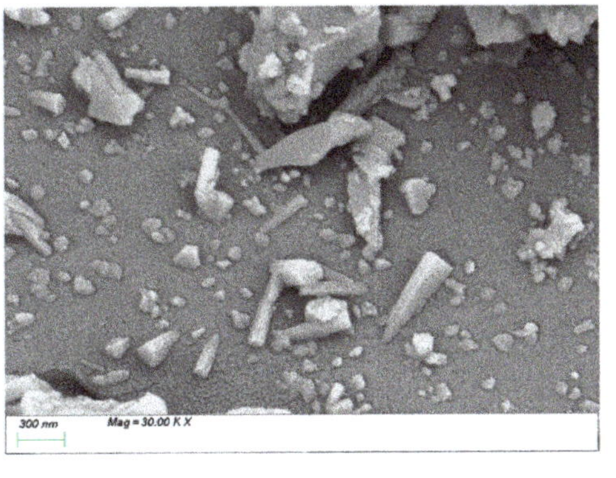

(b) Slag

Figure 2. SEM images of YRS and slag.

Table 2. Performance specification of fiber.

Fiber Type	Diameter (mm)	Density ($g.cm^{-3}$)	Length (mm)	Tensile Strength (MPa)	Elasticity (GPa)	Elongation (%)
ST	0.22	7.9	13	2800	210	5
PVA	0.04	1.3	12	1560	41	6.5
PE	0.02~0.03	0.9	12	3378	125	3

Table 3. The chemical composition of sodium silicate.

SiO_2/(%)	Na_2O/(%)	H_2O/(%)	Density/(g/cm^3)	Modulus	Beaume
30	13.5	56.5	1.51	2.3	50

2.1.2. Mixed Proportion

Prior to this research, experimental investigations were conducted to determine the optimal mix proportion parameters, as detailed in Section 3.1 "Main mix proportion parameters determination". Table 4 shows the mix proportion of slag–Yellow River sediment geopolymers with different fiber content. Before the experiment, the pre-weighed NaOH was thoroughly amalgamated with the sodium silicate. The mixture was subsequently concocted utilizing a uniaxial horizontal concrete mixer. Initially, the YRS along with the mineral admixtures were introduced into the mixer and agitated for a duration of 180 s. Following this, the fibers were incorporated into the mixer and the stirring continued for an additional 240 s. Subsequently, water was introduced into the mixer and the mixture was stirred for a further 180 s. Thereafter, the water glass was added and the mixture was stirred for 120 s. Immediately following this, the mixture was promptly poured into molds and then subjected to consolidation on a vibration table for a period ranging between 60 and 90 s. After 24 h of curing at room temperature, the specimens were demolded and stored in a curing chamber at 20 ± 2 °C and 95 ± 5% relative humidity until the designed curing ages [45,46].

Table 4. Mix proportion for slag–Yellow River sediment geopolymers with different fiber content.

No.	Sand	NaOH	SS	Slag	ST	PVA	PE	Water
REF	1.000	0.020	0.128	0.660	—	—	—	0.192
ST-0.5%	1.000	0.020	0.128	0.660	0.032	—	—	0.192
ST-1.0%	1.000	0.020	0.128	0.660	0.065	—	—	0.192
ST-1.5%	1.000	0.020	0.128	0.660	0.097	—	—	0.192
PVA-0.5%	1.000	0.020	0.128	0.660	—	0.006	—	0.192
PVA-1.0%	1.000	0.020	0.128	0.660	—	0.011	—	0.192
PVA-1.5%	1.000	0.020	0.128	0.660	—	0.017	—	0.192
PE-0.5%	1.000	0.020	0.128	0.660	—	—	0.004	0.192
PE-1.0%	1.000	0.020	0.128	0.660	—	—	0.007	0.192
PE-1.5%	1.000	0.020	0.128	0.660	—	—	0.011	0.192

2.2. *Experimental Method*

In this experiment, the effects of ST, PVA, and PE contents on the workability, mechanical elements, and characteristics of these products, alongside the microstructural properties of Yellow River sediment ecological cementitious materials, were analyzed through their compressive and splitting tensile strength, thermos gravimetric analysis, X-ray diffraction (XRD) analysis, and a scanning electron microscopy (SEM) test. Table 5 shows the grouping of the test, including the size and number of each test and specimen.

Table 5. Grouping of the workability, mechanical, and microstructural property tests.

Properties	Performance Index	Specimen Size	Quantity
Workability	Slump	—	—
	Slump flow	—	—
Strength	Compressive strength	100 mm × 100 mm × 100 mm	120
	Splitting tensile strength	100 mm × 100 mm × 100 mm	120
Characteristic products	Thermos gravimetric analysis	40 mm × 40 mm × 40 mm	30
	X-ray diffraction analysis	40 mm × 40 mm × 40 mm	30
Microstructural properties	Scanning electron microscopy	40 mm × 40 mm × 40 mm	30

2.2.1. Workability Test

The workability is a key index against which to measure the mixture. Whether it is easy to transport, pour, vibrate, and form, plays an extremely important role in ensuring the construction quality. Slump and expansion are two important evaluation indexes to measure the workability of mixtures. In this study, a slump bucket with a diameter of 50 mm at the top and 100 mm at the bottom and a height of 150 mm and a 60 cm × 60 cm flat plate were used to test the slump and slump flow of the mixture.

2.2.2. Strength Test

In order to fully evaluate the effect of fiber on the strength and toughness of matrix, the cube specimens with a side length of 100mm were used to test their compressive strength and splitting tensile strength, following the requirements outlined in GB/T 50081-2019 "Standard for Test Methods of Physical and Mechanical Properties of Concrete" [47]. The tension–compression ratio was calculated according to the test results.

2.2.3. X-Ray Diffraction Analysis

Using a Japanese Physical X-ray diffractometer, the sample was immersed in anhydrous ethanol for more than 7 days to terminate the reaction, then taken out of the sample and put it into a vacuum-drying oven for drying. After drying, the sample was ground, passed through a 200-mesh sieve, and put into a glass groove for testing. The sampling interval was 0.04° (2θ), the sampling speed was 1°/min, and the scanning angle range was 5°–70° (2θ).

2.2.4. Thermos Gravimetric Analysis

A ZCT-B simultaneous thermal analyzer (BEIJING JINGYI HITECHINSTRUMENT Co., Ltd., Beijing, China) was employed and the sample preparation method was the same as XRD. About 20mg of the sample was taken to be tested. The heating range was 30 °C~1000 °C, the heating rate was controlled at 10 °C/min, and the heating atmosphere was argon gas. The DTA curves for each group of samples were obtained.

2.2.5. Scanning Electron Microscopy Test

The microstructure of the cement paste samples was observed using a Sigma 300 field emission environmental scanning electron microscope (Carl Zeiss AG, Oberkochen, Germany). After curing to 28 d old, the samples were broken with pliers, immersed in anhydrous ethanol for more than 7 days to terminate the reaction, then place in a vacuum-drying oven for drying, and the test was conducted after the drying process was complete.

3. Experiment Results and Analysis

3.1. Main Mix Proportion Parameters Determination

Before conducting this study, it was essential to determine the appropriate mix proportion parameters. The water-to-binder ratio is a critical factor influencing the compressive strength of the matrix. An excessively low water-to-binder ratio can result in an overly concentrated alkali activator, a rapid reaction kinetics, and an uneven distribution of reaction products, which can ultimately degrade the material's mechanical performance. Conversely, an excessively high water-to-binder ratio would dilute the alkali activator, slow down the reaction rate (leading to insufficient product formation), and promote the evaporation of free water, which would increase porosity and reduce the compressive strength of the material. Consequently, this study first evaluated the effect of the water-to-binder ratio on the compressive strength of a slag–Yellow River sediment geopolymer. It should be noted that the total water content comprises two components: the water in the sodium silicate solution and the additional water added to the mixture. As shown in Figure 3a, the compressive strength of the matrix does not simply increase with a decreasing water-to-binder ratio; rather, it exhibits an optimal range of 0.38–0.40. Considering the subsequent need to incorporate fibers while maintaining the required workability, a water-to-binder ratio of 0.40 was selected for this study.

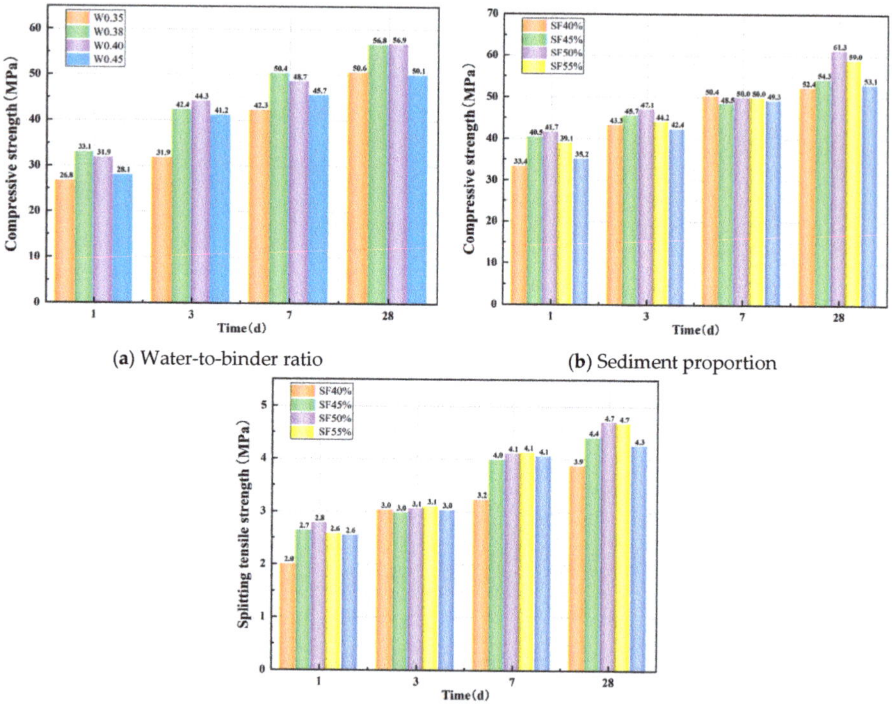

(**a**) Water-to-binder ratio

(**b**) Sediment proportion

(**c**) Sediment proportion

Figure 3. Effect of main mix proportion parameters on the strength of a slag–Yellow River sediment geopolymer.

Additionally, since the sediment from the Xixiayuan Reservoir used in this study has relatively fine particle sizes (with a most probable particle size of approximately 0.15 mm), its particle gradation and physicochemical composition differ significantly from conventional sand. Coupled with higher clay content, the influence of sediment proportion on matrix strength cannot be overlooked. A low proportion of Yellow River sediment offers limited economic benefits, while an excessively high proportion can lead to significant defects within the matrix. Furthermore, the clay fines tend to absorb a large amount of free water, which adversely affects the workability of the mixture. For these reasons, this study set the sediment proportion range at 40–60% and investigated its impact on matrix strength. As shown in Figure 3b,c, the matrix strength initially increases and then decreases as the sediment content rises. Both compressive strength and splitting tensile strength reach relatively high levels at a sediment proportion of 50%. Therefore, this study ultimately identified the optimal parameters as a water-to-binder ratio of 0.38 and a sediment proportion of 50%.

3.2. Workability

Figure 4 illustrates the effects of different mineral admixture replacement rates on the slump and slump flow of the mixture. Figure 4a specifically shows the influence of mineral admixture replacement rates on the slump. As observed in the figure, different mineral admixtures exhibit distinct effects on slump. With increasing steel fiber content, the slump of the mixture decreases significantly. When the steel fiber content reaches 2%, the slump drops sharply from 141 mm to 42 mm. PVAs and PEs display similar two-stage variation patterns. When the fiber content is within 1%, their impact on the fluidity of the paste is minimal. However, at a fiber content of 1%, the slumps for PVA-1% and PE-1% are 133 mm and 117 mm, respectively. Further increases in fiber content led to a marked decline in slump; at 2% fiber content, the slumps for PVA-2% and PE-2% plummet to 80 mm and 54 mm. Unlike slump, the slump-flow decreases significantly with increasing fiber content. As shown in Figure 4b, when the fiber content reaches 2%, the slump-flow values for STs, PVAs, and PEs drop sharply to 134 mm, 186 mm, and 146 mm, respectively.

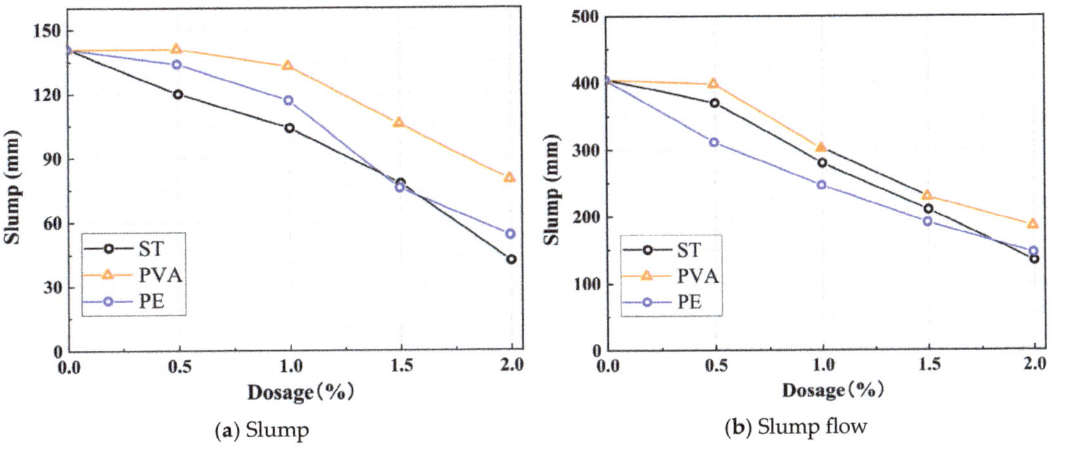

Figure 4. Effect of fiber on workability for (**a**) slump, and (**b**) slump flow.

This phenomenon occurs because steel fibers, due to their high density, increase the self-weight of the mixture. Additionally, their elevated surface roughness intensifies internal friction within the paste, leading to significant reductions in both slump and slump flow. In contrast, organic fibers, being lightweight and flexible, exhibit minimal impact on slump at low incorporation levels. However, their inherent water-absorbing properties deplete free water in the mixture, thereby reducing slump flow. When the fiber content exceeds a critical threshold, organic fibers form three-dimensional networks and agglomerates, which impede the flow of the paste, resulting in continuous declines in both slump and slump flow. In comparison, steel fibers exert the most pronounced adverse effects on workability. Among the organic fibers, PEs exhibit a more significant impact on performance degradation than PVAs.

3.3. Strength

3.3.1. Compressive Strength

Figure 5a–c illustrates the effects of STs, PVAs, and PEs on the compressive strength of the matrix. As shown in Figure 5a, the incorporation of an appropriate dosage of steel fibers significantly enhances the compressive strength of the matrix. After 1 day of curing, the compressive strength of the control group (without fibers) was 35.4 MPa. For ST0.5, ST1.0, and ST1.5, the compressive strengths were 44.5 MPa, 32.9 MPa, and 39.3 MPa, respectively. Except for ST1.0, which was slightly lower than the control group, ST0.5 and ST1.5 exhibited compressive strength increases of 25.68% and 10.93%, respectively, compared to the control group. As curing continued, all steel fiber-incorporated mixes demonstrated significantly higher compressive strengths than the control group. After 28 days of standard curing, the control group reached a compressive strength of 58.1 MPa, while ST0.5, ST1.0, and ST1.5 achieved compressive strengths of 71.5 MPa, 73.2 MPa, and 68.8 MPa, respectively, representing increases of 23.11%, 25.93%, and 18.43% over the control group.

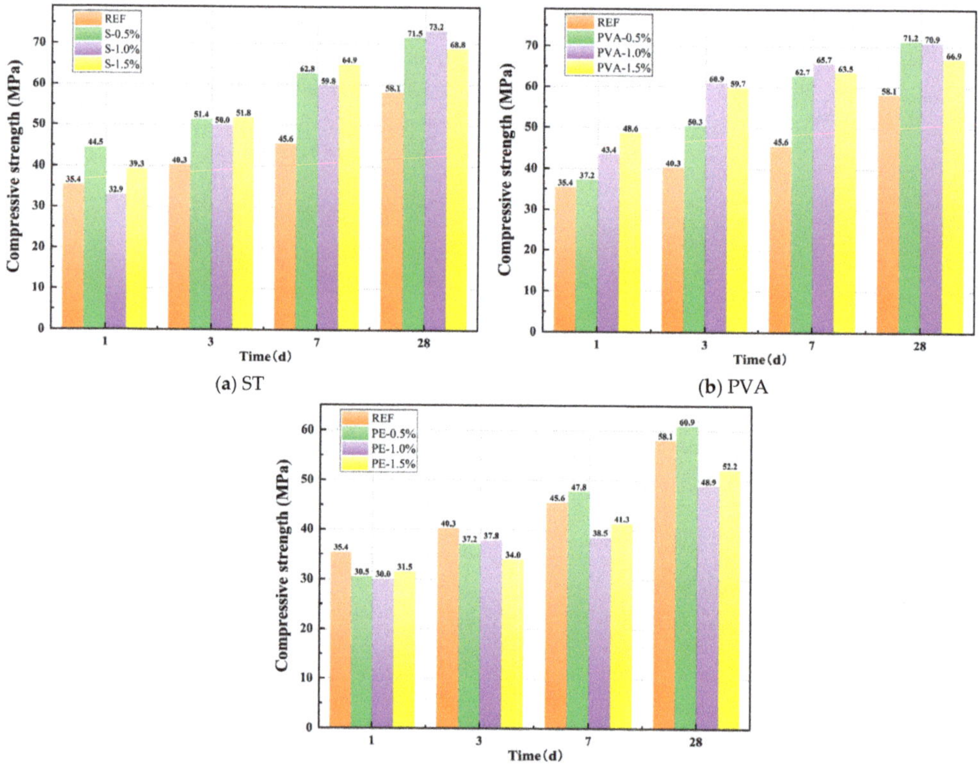

(**a**) ST (**b**) PVA (**c**) PE

Figure 5. Effect of fiber on compressive strength.

From Figure 5b, it can be observed that the incorporation of PVAs also significantly enhances the compressive strength of the matrix. After 1 day of curing, the compressive strengths of PVA0.5, PVA1.0, and PVA1.5 were 37.2 MPa, 43.4 MPa, and 48.6 MPa, respectively, representing increases of 5.06%, 22.71%, and 37.29% compared to the control group. As the curing age increased, all PVA-incorporated mixes maintained significantly higher compressive strengths than the control group. After 28 days of standard curing, the compressive strengths of PVA0.5, PVA1.0, and PVA1.5 reached 71.2 MPa, 70.9 MPa, and 66.9 MPa, respectively, corresponding to increases of 22.53%, 21.96%, and 15.06% over the control group.

From Figure 5c, it can be observed that, unlike steel fibers and PVAs, incorporating PEs slightly reduces the early-age compressive strength of the matrix. At 1 day of curing, the compressive strengths of PE0.5, PE1.0, and PE1.5 were 30.5 MPa, 30.0 MPa, and 31.5 MPa, respectively, representing decreases of 13.90%, 15.15%, and 11.05% compared to the control group. As the curing age continued to increase, the strength growth effect gradually became apparent under appropriate dosage ranges. After 28 days of standard curing, the compressive strengths of PE0.5, PE1.0, and PE1.5 reached 60.9 MPa, 48.9 MPa, and 52.2 MPa, respectively. While PE0.5 exhibited a 4.85% increase over the control group, PE1.0 and PE1.5 showed reductions of 15.81% and 10.25%, respectively.

In summary, both steel fibers and PVAs significantly enhance the early-age and later-age compressive strength of the control group, albeit through distinct mechanisms. Steel fibers primarily enhance compressive performance due to their high stiffness and tensile strength, while PVAs mitigate crack propagation through their flexibility and strong bonding properties. In contrast, the incorporation of PEs results in a slight reduction in early-age compressive strength for the sodium silicate-slag powder alkali-activated cementitious material. However, within appropriate dosage ranges, a gradual strength gain is observed at later ages, although the effectiveness of PEs remains less pronounced when compared to STs and PVAs.

3.3.2. Splitting Tensile Strength

Figure 6a–c illustrates the effects of STs, PVAs, and PEs on the compressive strength of geopolymer paste. From Figure 6a, it can be observed that incorporating steel fibers significantly enhances the splitting tensile strength of the matrix, with the effect becoming more pronounced as the fiber dosage increases. At 1 day of curing, the splitting tensile strength of the control group (without fibers) was 2.77 MPa. For ST0.5, ST1.0, and ST1.5, the splitting tensile strengths were 2.98MPa, 4.09MPa, and 5.20MPa, respectively, representing increases of 7.58%, 47.65%, and 87.73% compared to the control group. As the curing age increased, all steel fiber-incorporated mixes exhibited continuous growth in splitting tensile strength, consistently outperforming the control group. After 28 days of standard curing, the control group reached a splitting tensile strength of 4.06 MPa, while ST0.5, ST1.0, and ST1.5 achieved 5.63 MPa, 6.09 MPa, and 7.71 MPa, respectively, corresponding to improvements of 38.67%, 50.00%, and 89.90% over the control group.

From Figure 6b, it can be observed that incorporating PVAs also significantly improves the splitting tensile strength of the matrix. At 1 day of curing, the splitting tensile strengths of PVA0.5, PVA1.0, and PVA1.5 were 4.23 MPa, 4.59 MPa, and 4.85 MPa, respectively, representing increases of 52.71%, 65.70%, and 75.09% compared to the control group. As the curing age increased, all PVA-incorporated mixes maintained significantly higher splitting tensile strengths than the control group. After 28 days of standard curing, the splitting tensile strengths of PVA0.5, PVA1.0, and PVA1.5 reached 5.70 MPa, 6.51 MPa, and 6.67 MPa, respectively, corresponding to improvements of 40.39%, 60.34%, and 64.29% over the control group.

From Figure 6c, it can be observed that, unlike its effect on compressive strength, the incorporation of PEs significantly enhances the splitting tensile strength of the matrix. After 1 day of curing, the splitting tensile strengths of PE0.5, PE1.0, and PE1.5 were 3.71 MPa, 5.94 MPa, and 6.82 MPa, respectively, representing increases of 34.09%, 114.51%, and 146.08% compared to the control group. As the curing age progressed, the rate of splitting tensile strength growth in PE-incorporated mixes slowed, yet their strengths remained higher than the control group. After 28 days of standard curing, the splitting tensile strengths of PE0.5, PE1.0, and PE1.5 reached 4.57 MPa, 5.94 MPa, and 7.01 MPa, respectively, corresponding to improvements of 12.61%, 46.42%, and 72.61% over the control group. In summary, steel fibers, PVAs, and PEs all significantly enhance the tensile strength of the control group, with the toughening effect becoming more pronounced as the fiber dosage increases. Comparatively, PEs demonstrate the most significant enhancement in early-age tensile performance, while STs and PVAs exhibit more sustained and balanced strength improvements across all curing ages.

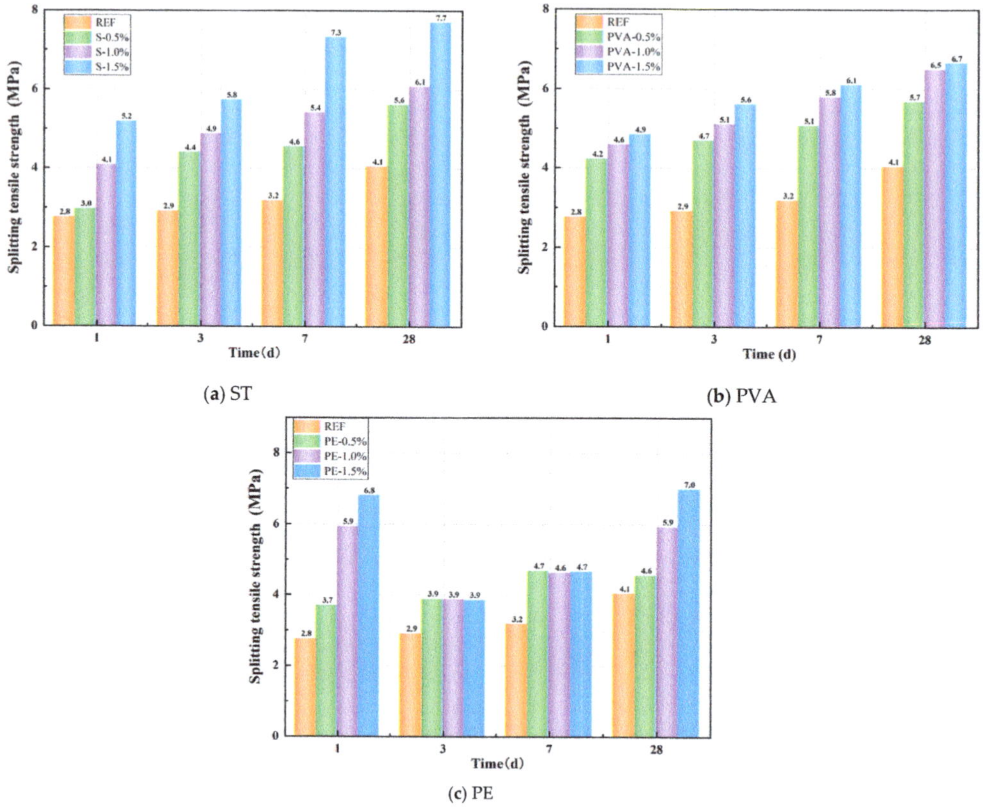

Figure 6. Effect of fiber on splitting tensile strength.

3.3.3. Tension–Compression Ratio

The tension–compression ratio is a critical parameter for evaluating the brittle characteristics of concrete, reflecting the material's mechanical behavior under both tensile and compressive stress conditions. A lower tension–compression ratio indicates that concrete is more susceptible to brittle failure under tensile stress, exhibiting poor crack resistance. In contrast, a higher tension–compression ratio suggests improved material toughness, which helps control crack formation and delays crack propagation, thereby enhancing the overall performance of concrete structures. Variations in the tension–compression ratio not only influence the load-bearing capacity of structures but also have significant implications for their durability and service life. Therefore, optimizing the tension–compression ratio is essential for ensuring structural integrity and long-term durability in engineering applications.

Figure 7a–c illustrates the effects of STs, PVAs, and PEs on the tension–compression ratio of a geopolymer paste. From Figure 7a, it can be observed that all PVA-incorporated mixes exhibit significantly higher tension–compression ratios than the control group, with the ratio increasing more markedly as fiber dosage rises. At 1 day of curing, the tension–compression ratio of ST0.5 was 14.40% lower than the control group, while ST1.0 and ST1.5 showed increases of 58.73% and 69.23%, respectively. As the curing age increased, all steel fiber-incorporated mixes maintained significantly higher tension–compression ratios than the control group. After 28 days of standard curing, the tension–compression ratios of ST0.5, ST1.0, and ST1.5 increased by 12.64%, 19.11%, and 60.35%, respectively, compared to the control group.

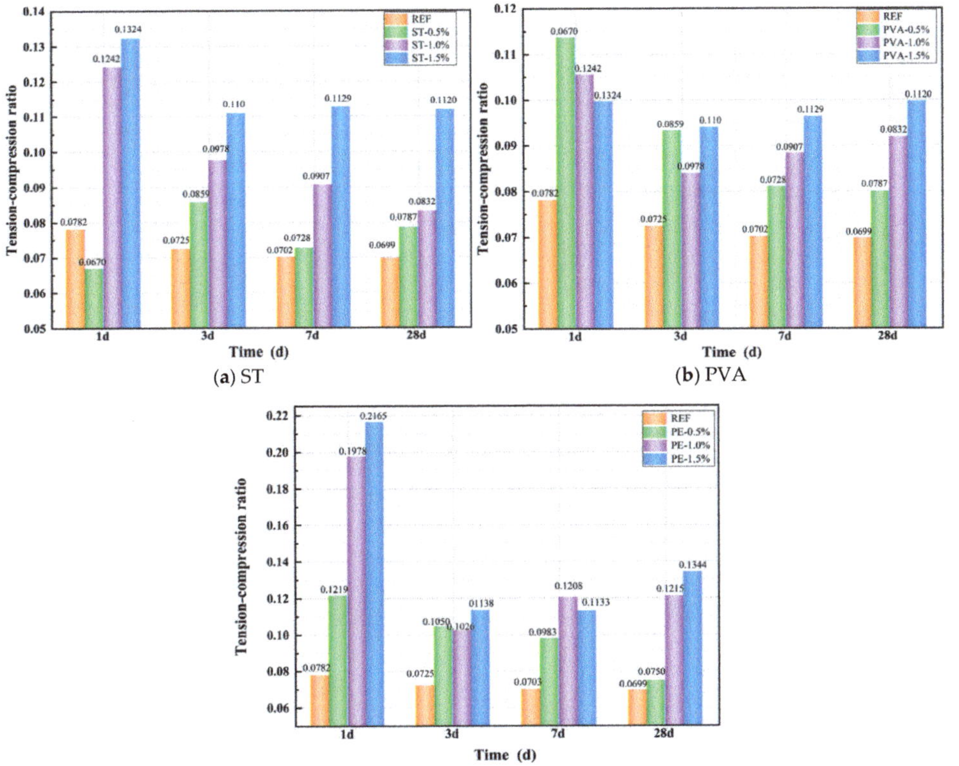

(a) ST (b) PVA

(c) PE

Figure 7. Effect of fiber on tension–compression ratio.

From Figure 7b, it can be observed that all PVA-incorporated mixes exhibit significantly higher tension–compression ratios than the control group. However, unlike the trend observed with steel fibers, the early-age tension–compression ratio decreases as fiber content increases. After 1 day of curing, the tension–compression ratios of PVA0.5, PVA1.0, and PVA1.5 were 45.36%, 35.04%, and 27.53% higher than the control group, respectively. As the curing age progressed, the influence of higher fiber content on the tension–compression ratio became more pronounced. After 28 days of standard curing, the tension–compression ratios of PVA0.5, PVA1.0, and PVA1.5 increased by 14.58%, 31.48%, and 42.79%, respectively, compared to the control group.

From Figure 7c, it can be observed that all PE-incorporated mixes exhibit significantly higher tension–compression ratios than the control group, with the ratio increasing more markedly as fiber dosage rises. At 1 day of curing, the tension–compression ratios of PE0.5, PE1.0, and PE1.5 were 34.09%, 114.51%, and 146.08% higher than the control group, respectively. As the curing age progressed, the tension–compression ratio exhibited a declining trend due to the slower growth of splitting tensile strength and the continuous improvement in compressive strength in the PE-incorporated mixes. Nevertheless, at all curing ages, the tension–compression ratios of PE-incorporated specimens remained significantly higher than those of the control group. After 28 days of standard curing, the tension–compression ratios of PE0.5, PE1.0, and PE1.5 increased by 7.41%, 73.91%, and 92.32%, respectively, compared to the control group.

3.4. Five-Dimensional Evaluation

Based on the experimental results, this study selected five key indicators—compressive strength, splitting tensile strength, tension–compression ratio, slump, and slump flow—and employed a five-dimensional evaluation method [48] for comparative analysis to determine the optimal dosage of STs, PVAs, and PEs. The multidimensional evaluation results of slag–Yellow River sediment geopolymers under different fiber dosages are shown in Figure 8. From Figure 8a, it can be observed that incorporating 1.0% steel fibers provides a balanced enhancement of matrix strength: compressive strength, splitting tensile strength, and tension–compression ratio increased by 25.93%, 50.00%, and 19.11%, respectively, compared to the control group. In contrast, 1.5% steel fibers yielded the most significant improvement in tensile performance, with increases of 18.43%, 89.90%, and 60.35% in compressive strength, splitting tensile strength, and tension–compression ratio, respectively. However, at this dosage (1.5%), workability declined significantly. From Figure 8b, similar to steel fibers, 1.0% of PVAs achieved balanced strength improvements as the compressive strength, splitting tensile strength, and tension–compression ratio increased by 21.96%, 60.34%, and 31.48%, respectively. Meanwhile, 1.5% PVAs maximized tensile strength enhancement, with increases of 15.06%, 64.29%, and 42.79% in the respective metrics. Nevertheless, workability also deteriorated markedly at this higher dosage. From Figure 8c, PEs exhibited uneven strength enhancement. Incorporating 0.5% PEs delivered the best compressive strength improvement but limited tensile performance gains, while 1.5% PEs maximized tensile strength enhancement but adversely affected workability and compressive strength. In summary, to ensure satisfactory workability and balance performance metrics, 1.0% dosage is recommended for both STs and PVAs.

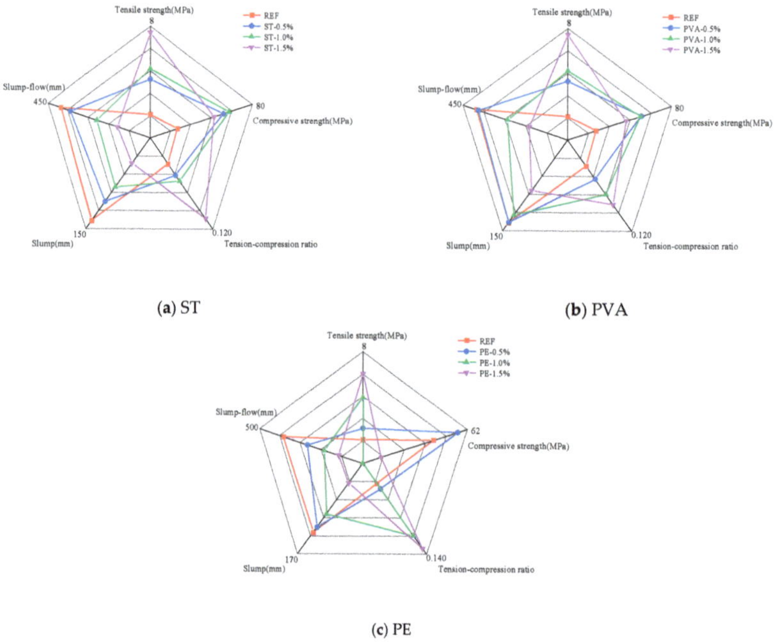

Figure 8. Five-dimensional evaluation diagram under different ST, PVA, and PE content.

4. Influence Mechanism Analysis

4.1. Reaction Products

Due to the complex composition of the silted sediment, paste samples excluding sediment were prepared for XRD and TGA tests to avoid interference from impurities and minerals in the sediment. Figure 9 shows the TGA curves under different ST, PVA, and PE contents. As observed in the figure, the TGA curves exhibit two primary thermal decomposition peaks: one in the 70–100 °C temperature range and another in the 700–850 °C range.

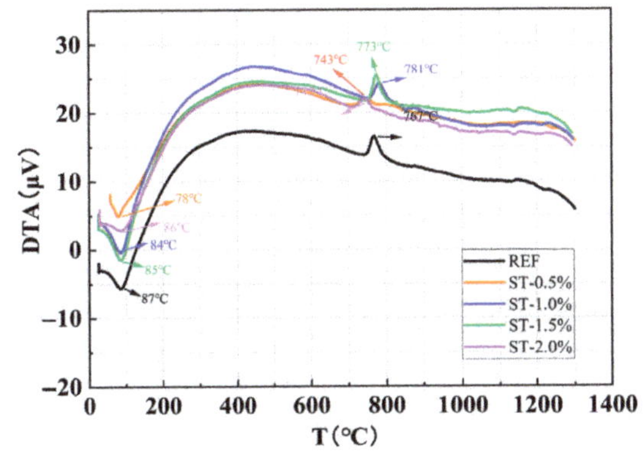

(**a**) ST

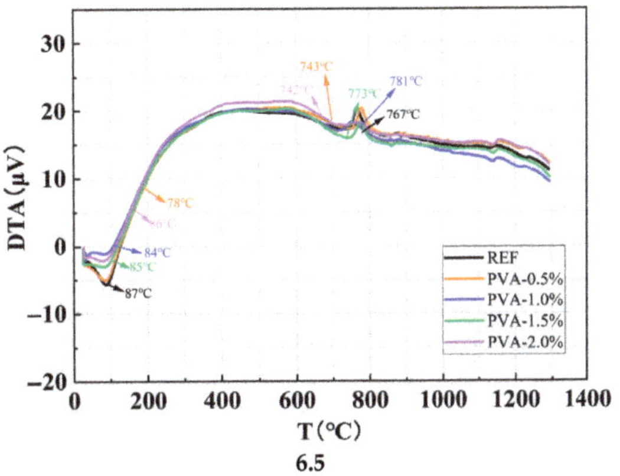

(**b**) PVE

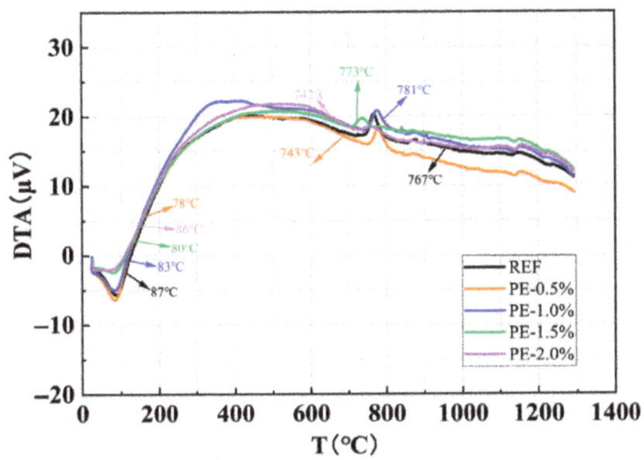

(**c**) PE

Figure 9. Comparative study of thermal analysis curves under different ST, PVA, and PE content.

From Figure 9a, it can be observed that the low-temperature decomposition peaks of the control group, ST0.5, ST1.0, ST1.5, and ST2.0 are located at 87 °C, 78 °C, 84 °C, 85 °C, and 86 °C, respectively, while their high-temperature decomposition peaks are positioned at 767 °C, 743 °C, 781 °C, 773 °C, and 742 °C, respectively. When the steel fiber content reaches 2.0%, the intensity of the high-temperature decomposition peak decreases, indicating that excessive steel fibers reduce the number of characteristic gel products formed. However, the type of reaction products remains unchanged.

From Figure 9b, it can be observed that as the PVA content increases, the low-temperature decomposition peaks of the control group, PVA0.5, PVA1.0, PVA1.5, and PVA2.0 are located at 87 °C, 85 °C, 84 °C, 81 °C, and 79 °C, respectively, while their high-temperature decomposition peaks occur at 767 °C, 777 °C, 756 °C, 774 °C, and 777 °C, respectively. The high-temperature decomposition peak intensities for PVA1.5 and PVA2.0 are lower than those of other mixes, and the low-temperature decomposition peak intensities for PVA1.0, PVA1.5, and PVA2.0 are also reduced. These results indicate that while PVAs do not alter the types of reaction products, the hydroxyl (-OH) groups on PVAs impart remarkable hydrophilicity (with a typical contact angle of <30°) and they influence the distribution of free water and the quantity of characteristic gel products formed. Moreover, at equivalent dosages, the impact of PVAs on these properties is more pronounced compared to steel fibers.

From Figure 9c, it can be observed that as the PE content increases, the low-temperature decomposition peaks of the control group, PE0.5, PE1.0, PE1.5, and PE2.0 are located at 87 °C, 84 °C, 83 °C, 80 °C, and 72 °C, respectively, while their high-temperature decomposition peaks occur at 767 °C, 778 °C, 775 °C, 736 °C, and 757 °C, respectively. Both the low-temperature and high-temperature decomposition peak intensities for PE1.5 and PE2.0 are lower than those of other mixes. These results indicate that while PEs do not alter the types of reaction products, they significantly influence the distribution of free water and the quantity of characteristic gel products formed. Furthermore, at equivalent dosages, the impact of PEs on these properties is more pronounced compared to steel fibers.

In summary, the incorporation of steel fibers, PVAs, and PEs does not alter the types of characteristic reaction products in alkali-activated cementitious materials. However, these fibers significantly influence the formation of gel-like products and the distribution of free water, thereby impacting both the quantity and the thermal decomposition behavior of the reaction products. The effects of PVAs and PEs are notably more pronounced compared to steel fibers, highlighting their more substantial influence on the microstructural properties of the matrix.

Figure 10 shows the XRD patterns of a matrix under different ST, PVA, and PE content. Respectively. As can be seen from the figures, whether in Figure 10a,b or Figure 10c, the main reaction products of the sample are C-A-S-H. The results indicate that fibers influence the matrix product content without inducing the formation of new reaction products. As fiber content increases, the peak intensity of the C-A-S-H and C-S-H phases decreases. While all fiber types affect the formation of gel-like products, their mechanisms of action vary. With increasing steel fiber content, excessive fiber aggregation within the matrix leads to the formation of additional interfacial transition zones (ITZs), which disrupt the spatial distribution of the reactions, resulting in the non-uniform dispersion of primary gel-like products and incomplete polymerization. In contrast, PVAs and PEs influence the reactions through distinct mechanisms. PVAs, due to their hydrophilic nature, adsorb free water, reducing the available water for the "dissolution-depolymerization-condensation" reaction and thereby limiting the formation of characteristic gel products. Additionally, PVAs alter the pore structure of the matrix, further modifying the spatial distribution of the reactions. Conversely, PEs, owing to their hydrophobicity, impede water migration and distribution, disrupting the homogeneity of the reaction and consequently reducing the quantity of gel products formed. These observations align with thermal analysis results which indicate that fibers primarily act as physical reinforcements within the system. Their chemically inert nature ensures minimal interaction with the alkali-activation reaction process. As a result, fibers influence reaction uniformity and the quantity of reaction products formed primarily through physical mechanisms, without altering the types of reaction products.

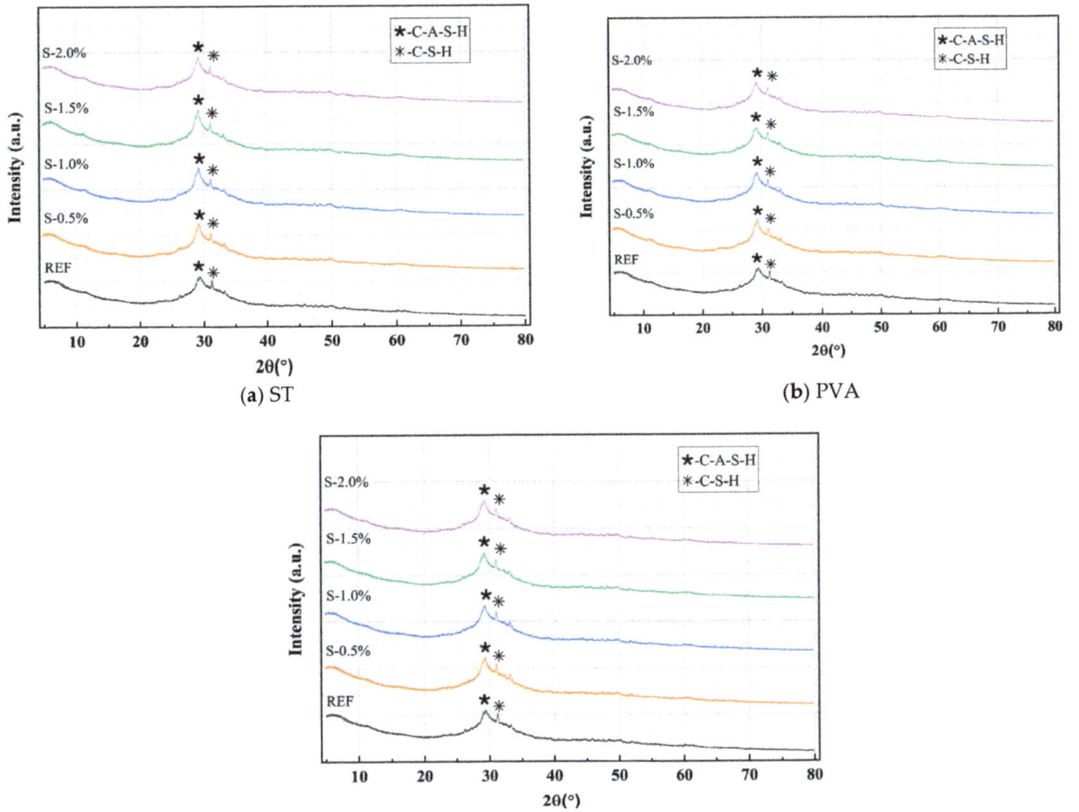

(**a**) ST (**b**) PVA (**c**) PE

Figure 10. Comparative study of XRD patterns under different ST, PVA, and PE content.

4.2. Matrix Microstructure

Figure 11 presents the SEM test results of the control group. As shown in Figure 11a, after 28 days of curing, numerous plate-like C-(A)-S-H gels were formed within the matrix. These gel products exhibited sufficient growth [49] and effectively filled the matrix voids, forming a uniform, dense, continuous, and intact microstructure, thereby contributing to the high compressive strength. From Figure 11b, it can be observed that most slag particles were completely reacted, with tightly packed, well-developed gel products displaying a higher degree of polymerization. This resulted in a dense matrix structure [50]. This phenomenon is attributed to the appropriate alkali content, which enhances the dissolution rate of slag particles, accelerates the reaction process, increases the reaction extent, and ultimately promotes the formation of C-A-S-H and C-S-H gels.

Figure 12 presents the SEM results of the sample incorporating 0.5% STs. As shown in Figure 12a, the steel fibers exhibit tight physical bonding with the matrix structure, with a distinct interfacial transition zone characterized by strong mechanical interlocking. The high surface roughness of the steel fibers enhances mechanical anchorage with the matrix, while their high modulus, stiffness, and large diameter effectively improve the strength and toughness of the matrix. From Figure 12b, it can be observed that the primary reaction products within the matrix remain plate-like C-(A)-S-H gels, indicating that the incorporation of steel fibers does not alter the types or microstructural morphology of the characteristic reaction products. However, the high modulus of steel fibers enhances the material's overall mechanical performance by inhibiting crack propagation through fiber-bridging effects.

(a) (b)

Figure 11. Microscopic image of REF. (**a**) Microstructure, (**b**) characteristic product.

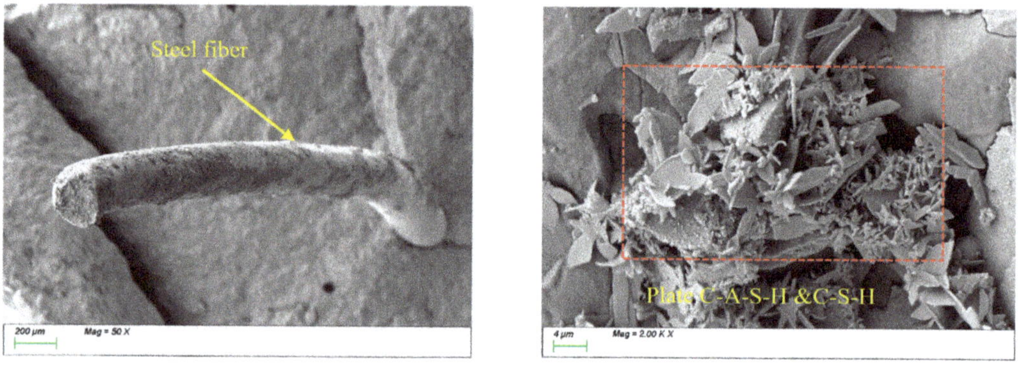

(a) (b)

Figure 12. Microscopic image of S0.5. (**a**) Steel fiber distribution, (**b**) characteristic product.

Figure 13 presents the SEM results of the sample incorporating 1.5% STs. As shown in Figure 13a, the randomly oriented steel fibers form tightly interlocked networks, effectively enhancing the tensile performance of the matrix. However, the non-uniform distribution of fibers at this high dosage introduces voids at ITZs, compromising the matrix compactness. Consequently, compared to the 0.5% STs sample, the 1.5% STs specimen exhibits higher tensile strength but reduced compressive strength due to increased interfacial defects. From Figure 13b, the dominant reaction products within the matrix remain plate-like C-(A)-S-H gels, confirming that steel fiber dosage does not alter the types or microstructural morphology of characteristic products. Nevertheless, the voids and defects at ITZs formed by excessive steel fibers may act as stress concentration points, adversely affecting the material's overall performance.

Figure 14 presents the SEM results of the sample incorporating 0.5% PVAs. As shown in Figure 14a, the randomly oriented PVAs interconnect within the matrix, forming a three-dimensional spatial network structure. The flexibility and high ductility of PVAs enable them to effectively bridge microcracks at the microscale, significantly enhancing the material's toughness. From Figure 14b, it can be observed that the internal matrix products are dominated by C-(A)-S-H gels, indicating that the incorporation of PVAs does not alter the types or microstructural morphology of characteristic reaction products. However, through fiber bridging effects, they suppress crack propagation, thereby improving the overall mechanical performance of the material.

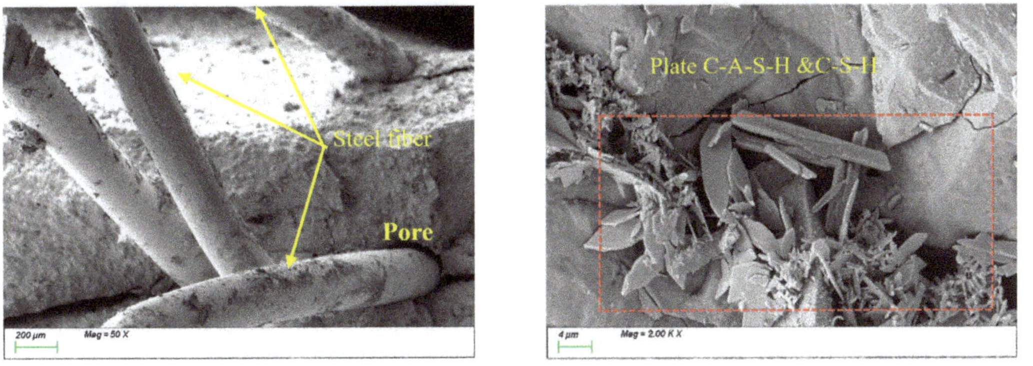

(**a**) (**b**)

Figure 13. Microscopic image of S1.5. (**a**) Steel fiber distribution, (**b**) characteristic product.

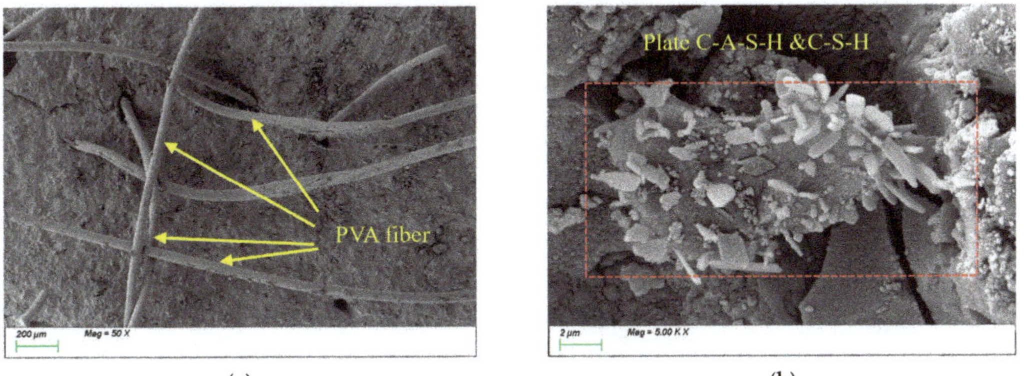

(**a**) (**b**)

Figure 14. Microscopic image of PVA0.5. (**a**) Polyvinyl alcohol fiber distribution, (**b**) characteristic product.

Figure 15 presents the SEM results of the sample incorporating 1.5% PVAs. As shown in Figure 15a, the increased PVA content results in a dense network of randomly distributed, closely interlaced fibers, significantly enhancing the tensile performance of the matrix. During fiber stretching, channels formed within the matrix (as shown in Figure 15b), which consumed substantial energy and contributed to the development of the matrix's integrity and tensile strength. However, the non-uniform distribution of high PVAs content introduces voids at ITZs (as shown in Figure 15a), reducing the matrix compactness and thereby adversely affecting compressive strength. In summary, the incorporation of PVAs markedly improves the toughness and tensile performance of sodium silicate-slag powder alkali-activated cementitious materials. Nevertheless, the fiber dosage must be optimized to avoid interfacial defects that compromise the material's compactness and compressive strength.

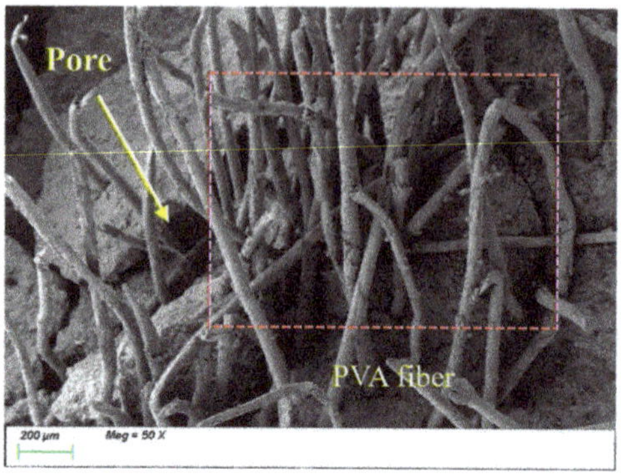

(**a**)

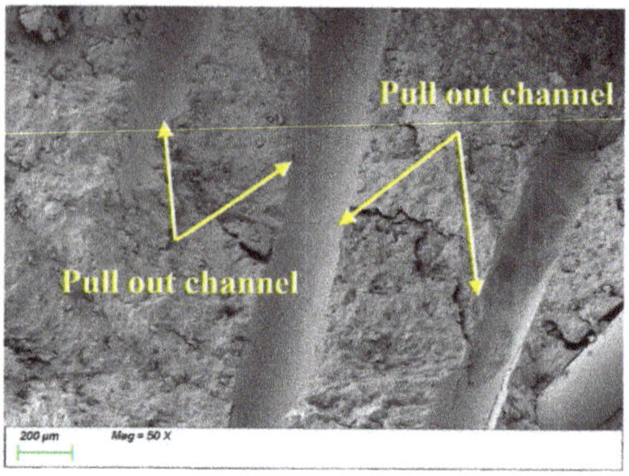

(**b**)

Figure 15. Microscopic image of PVA1.5. (**a**) Polyvinyl alcohol fiber distribution, (**b**) channels formed during fiber stretching.

Figure 16 presents the SEM results of the sample incorporating 0.5% polyethylene fibers (PEs). As shown in Figure 16a, compared to steel fibers, PEs exhibit lower density and a higher number of fibers per unit volume, facilitating spatial structural interlocking within the matrix and providing a certain reinforcing effect. From Figure 16b, it can be observed that the C-S-H and C-A-S-H gels within the matrix display honeycomb-like growth with a lower degree of polymerization and a relatively loose structure. This is likely attributed to the hydrophobicity and low surface energy of PEs, which weaken interfacial bonding with the matrix and result in inferior distribution uniformity compared to PVAs. Additionally, the incorporation of PEs may impede water migration and distribution, compromising reaction homogeneity and thereby restricting the quantity, distribution, and polymerization degree of characteristic gel products.

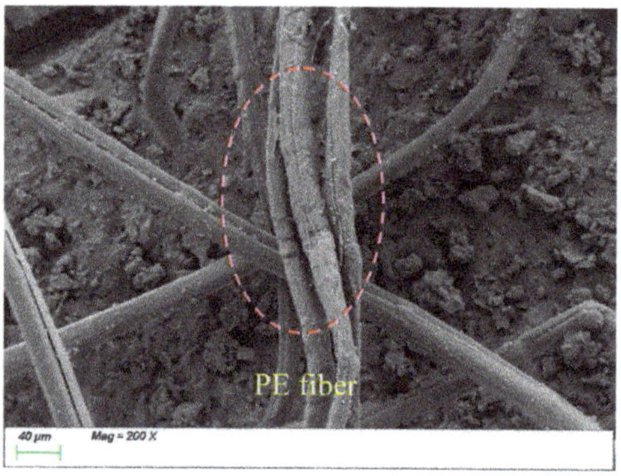

(**a**)

(**b**)

Figure 16. Microscopic image of PE0.5. (**a**) Polyethylene fiber distribution, (**b**) characteristic product.

Figure 17 presents the SEM results of the sample incorporating 1.5% polyethylene fibers (PEs). Compared to Figure 17a, the increased fiber content results in a significantly higher number of fibers within the matrix. The densely randomly oriented PEs form a three-dimensional network structure through extensive interlocking, which enhances the tensile strength of the matrix. However, the figure also reveals that excessive fiber content leads to non-uniform fiber distribution and the presence of large pores within the matrix, adversely affecting its compactness. From Figure 17b, it can be observed that the C-S-H and

C-A-S-H gels within the matrix still exhibit honeycomb-like growth with a lower degree of polymerization and a loose gel structure. This results in weak bonding between the gels and PEs, which is likely one of the primary reasons why the compressive strength of the 1.5% PEs sample is lower than both the control group and the 0.5% PEs sample. In summary, the bridging effect of PEs primarily operates at the macroscopic scale. While high-tensile, high-elastic-modulus PEs improve the tensile performance of the matrix, their microscale reinforcement effect remains limited due to insufficient interfacial bonding and a compromised gel structure.

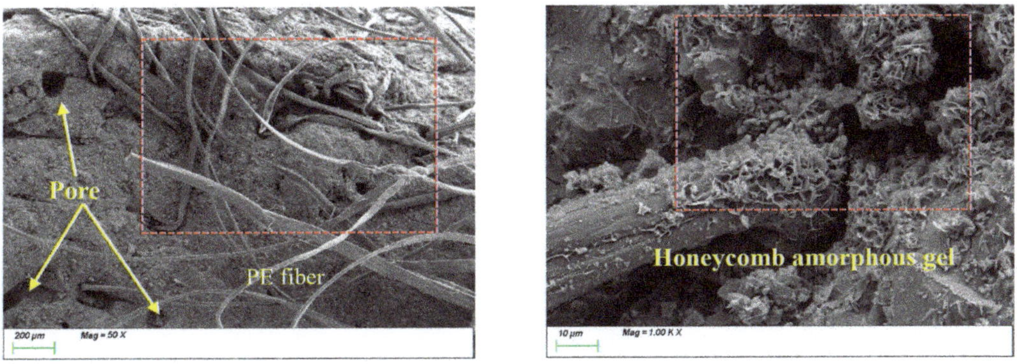

(**a**) (**b**)

Figure 17. Microscopic image of PE1.5. (**a**) Polyethylene fiber distribution, (**b**) characteristic product.

5. Conclusions

This study evaluated the impact of ST, PVA and PE on several critical properties of slag–Yellow River sediment geopolymers, including the workability, compressive strength, splitting tensile strength, characteristic products, and matrix microstructure. The following are the key findings of the study:

(1) As the fiber content increases, the workability (slump and spread) of the mixture significantly decreases. STs have the most pronounced effect on workability, with ST2.0 slump and spread reduced to 29.8% and 33.1% of the initial values, respectively. PVAs and PEs have a minor impact on slump at low dosages but significantly reduce workability when the dosage exceeds 1.0%. PVA2.0 slump and spread decrease to 56.7% and 45.9% of the initial values, respectively, while PE2.0 slump and spread decrease to 38.3% and 36.0% of the initial values, respectively. The addition of fibers increases internal friction and self-weight, with the high density of STs further hindering paste fluidity.

(2) An appropriate amount of STs and PVAs can significantly enhance the compressive strength and splitting tensile strength of the matrix. The 28-day compressive strength of ST1.0 and PVA1.0 increased by 25.93% and 21.96%, respectively, compared to the control group. The 28-day splitting tensile strength of ST1.5 and PVA1.5 increased by 89.90% and 64.29%, respectively, compared to the control group. The 28-day tensile-to-compressive ratio of ST1.5 and PVA1.5 increased by 60.35% and 42.79%, respectively, compared to the control group. In contrast, PEs primarily contribute to toughness, with PE1.5′s 28-day splitting tensile strength increasing by 72.61% and the 28-day tensile-to-compressive ratio increasing by 92.32% compared to the control group.

(3) With an increase in age, all fiber-reinforced specimens (containing STs, PVAs or PEs) exhibited significantly higher increasing rates in the tensile-to-compressive strength ratio compared to the reference group, and the ratio enhancement became more

pronounced with higher fiber content; after 28 days of standard curing the ST-0.5, ST-1.0, and ST-1.5 mixtures showed 12.64%, 19.11%, and 60.35% improvements in the ratio, respectively, while the PVA-0.5, PVA-1.0, and PVA-1.5 mixtures achieved 14.58%, 31.48%, and 42.79% enhancements, respectively, and the PE-0.5, PE-1.0, and PE-1.5 mixtures attained 7.41%, 73.91%, and 92.32% increases, respectively, which conclusively verified the effective development of fiber bridging effects in the cementitious composites.

(4) STs, PVAs, and PEs do not alter the types of characteristic products in geopolymer cementitious materials, but they affect the generation of gel-like products and the distribution of free water to varying degrees. PVAs and PEs have a more significant impact than STs, and excessive fiber content can affect the generation of characteristic gel products and their thermal decomposition behavior.

(5) STs enhance the compressive performance of the matrix through their high modulus and stiffness and inhibit crack propagation through fiber bridging. PVAs suppress crack propagation through their flexibility and good bonding properties, while PEs improve the tensile performance of the matrix through their high tensile strength and elastic modulus. Although their reinforcing effect on the matrix at the microscopic scale is relatively weak, the interfacial properties between PEs and the cement matrix could be optimized through surface treatments (e.g., chemical modification or coating applications) in future studies, thereby enhancing fiber–matrix interactions.

Author Contributions: Conceptualization, G.Z., E.J. and C.Y.; methodology, G.Z., E.J. and K.L.; validation, G.Z. and H.S.; formal analysis, G.Z., H.S. and C.C.; investigation, G.Z., C.C. and H.S.; data curation, G.Z., C.Y. and H.S.; writing—original draft preparation, G.Z.; writing—review and editing, G.Z., E.J. and K.L.; funding acquisition, G.Z. All authors have read and agreed to the published version of the manuscript.

Funding: This research was funded by the Science and Technology Development Fund of the Yellow River Institute of Hydraulic Research (202112), the Natural Science Foundation of Henan (grant number 242300421462), the Excellent Young Talents Project of Yellow River Conservancy Commission (No. HQK-202310), the Major Science and Technology Project of Henan Province (No. 231100320100), and Henan Provincial Key R&D and Promotion Special Project (Science and Technology Tackling) (242102321156).

Data Availability Statement: The original contributions presented in the study are included in the article. Further inquiries can be directed to the corresponding author.

Conflicts of Interest: The authors declare no conflicts of interest.

Nomenclature

STs	Steel fibers	nm	Nanometer
YRS	Yellow river sediment	#	Number
C-S-H	Calcium silicate hydrate	P	Pressure
C-A-S-H	Calcium aluminum silicate hydrate	XRD	X-ray Diffraction
ITZs	Interfacial Transition Zones	m	Meter
N	Newton	SiO_2	Silicon Dioxide
PVAs	Polyvinyl Alcohol fibers	g	Gram
PEs	Polyethylene fibers	g cm^{-3}	Grams per Cubic Centimeter
MPa	Megapascal	SO_4^{2-}	Sulfate lon
OPC	Ordinary Portland cement	N	Newton
TGA	Thermogravimetric Analysis	SiO_2	Silicon dioxide
SEM	Scanning electron microscope	SG	Specific gravity

mm	Millimeter	SS	Sodium Silicate
m^2	Square meters	μW	Microwatt
$m^2\ kg^{-1}$	Square meters per kilogram	NaOH	Sodium Hydroxide
min	Minute	%	Percentage

References

1. Yang, Z.X.; Zeng, L.Y. Research on the Current Situation of Mining Management of National Sandy and Stony soil Mines. *Nat. Resour. Econ. China* **2020**, *33*, 44–50. [CrossRef]
2. Wei, J.; Cen, K. Empirical assessing cement CO_2 emissions based on China's economic and social development during 2001–2030. *Sci. Total Environ.* **2019**, *653*, 200–211. [CrossRef] [PubMed]
3. Liu, S.; Liu, C.Y.; Hao, Y.F.; Zhang, Y.; Chen, L.; Li, Z. Experimental investigation of engineered geopolymer composite for structural strengthening against blast loads. *Def. Technol.* **2024**, *32*, 496–509. [CrossRef]
4. Yang, J.; Wang, Z.Q.; He, X.Y.; Su, Y.; Tang, Y.Z.; Qi, H.H.; Yang, C.; Xiong, G.Q. Using superabsorbent polymer to mitigate the fast setting and high autogenous shrinkage of carbide slag and sodium silicate activated ultrafine GGBS based composites. *Sustain. Chem. Pharm.* **2024**, *39*, 101550. [CrossRef]
5. Cao, B.S.; Li, Y.; Li, P.P. Synergistic Effect of Blended Precursors and Silica Fume on Strength and High Temperature Resistance of Geopolymer. *Materials* **2024**, *17*, 2975. [CrossRef]
6. Yang, N.; Xuan, Q.D.; Fu, Y.; Ma, X.; Lei, D.Y.; Niu, J.L.; Dai, J.G. Phosphate activated geopolymer-based coating with high temperature resistance for sub-ambient radiative cooling. *Sustain. Cities Soc.* **2024**, *100*, 104992. [CrossRef]
7. Wang, Z.K.; Fu, C.H.; Wang, K.; Zhao, J.; Shumuye, E.D.; Yang, Z.H. Effect of geopolymer concrete cover on improving tensile and transverse shear behaviors of BFRP bars after exposure to high temperature. *Case Stud. Constr. Mater.* **2024**, *20*, e02862. [CrossRef]
8. Chen, Z.M.; Liu, H.; Zhu, P.H.; Li, H.C.; Ge, T.Z.; Yang, L.; Chen, C.H.; Dong, Y.L. Effect of Curing Mechanism on Sulfuric Acid Corrosion Resistance of Geopolymer Recycled Aggregate Concrete. *Ksce J. Civ. Eng.* **2024**, *28*, 1173–1182. [CrossRef]
9. Zhou, Y.W.; Yu, Y.; Guo, W.H.; Xing, F.; Guo, M.H. Development of inorganic anticorrosive coatings for steel bars: Corrosion resistance testing and design. *Cem. Concr. Compos.* **2024**, *152*, 105612. [CrossRef]
10. Cui, L.; Xiang, T.; Hu, B.; Lv, Y.; Rong, H.; Liu, D.E.; Zhang, S.; Guo, M.; Lv, Z.; Chen, D. Design of monolithic super hydrophobic concrete with excellent anti-corrosion and self-cleaning properties. *Colloids Surf. A Physicochem. Eng. Asp.* **2024**, *685*, 133345. [CrossRef]
11. Gopalakrishnan, R.; Chinnaraju, K. Durability of ambient cured alumina silicate concrete based on slag/fly ash blends against sulfate environment. *Constr. Build. Mater.* **2019**, *204*, 70–83. [CrossRef]
12. Ahmad, M.R.; Chen, B.; Shah, S.F.A. Influence of different admixtures on the mechanical and durability properties of one-part alkali-activated mortars. *Constr. Build. Mater.* **2020**, *265*, 120320. [CrossRef]
13. Alzeer, M.I.M.; MacKenzie, K.J.D. Chapter 5—Fiber composites of inorganic polymers (geopolymers) reinforced with natural fibers. In *Composite Materials*; Low, I.-M., Dong, Y., Eds.; Elsevier: Amsterdam, The Netherlands, 2021; pp. 117–147. [CrossRef]
14. Krishna, R.S.; Shaikh, F.; Mishra, J.; Lazorenko, G.; Kasprzhitskii, A. Mine tailings-based geopolymers: Properties, applications and industrial prospects. *Ceram. Int.* **2021**, *47*, 17826–17843. [CrossRef]
15. Nuaklong, P.; Worawatnalunart, P.; Jongvivatsakul, P.; Tangaramvong, S.; Pothisiri, T.; Likitlersuang, S. Pre- and post-fire mechanical performances of high calcium fly ash geopolymer concrete containing granite waste. *J. Build. Eng.* **2021**, *44*, 103265. [CrossRef]
16. Figiela, B.; Bąk, A.; Hebda, M.; Korniejenko, K. Eco-friendly production of foamed geopolymers based on mine waste. *J. Achiev. Mater. Manuf. Eng.* **2023**, *121*, 341–349. [CrossRef]
17. Dong, C.H.; Li, T.; Zhang, Y.M.; Liu, J. Damage process and performance of PVA fiber-reinforced alkali-activated slag mortar plate under bending. *J. Southeast Univ. (Engl. Ed.)* **2018**, *34*, 229–236. [CrossRef]
18. Zhang, S.Z.; He, S.; Ghiassi, B.; Breugel, K.V.; Ye, G. Interface bonding properties of polyvinyl alcohol (PVA) fiber in alkali-activated slag/fly ash. *Cem. Concr. Res.* **2023**, *173*, 107308. [CrossRef]
19. Sarker, P.K.; Haque, R.; Ramgolam, K.V. Fracture behaviour of heat cured fly ash based geopolymer concrete. *Mater. Des.* **2013**, *44*, 580–586. [CrossRef]
20. Pan, Z.; Sanjayan, J.G.; Rangan, B.V. Fracture properties of geopolymer paste and concrete. *Mag. Concr.* **2011**, *63*, 763–771. [CrossRef]
21. Thomas, R.J.; Peethamparan, S. Alkali-activated concrete: Engineering propertie sand stress strain behavior. *Constr. Build. Mater.* **2015**, *93*, 49–56. [CrossRef]

22. Atis, C.D.; Bilim, C.; Celik, O.; Karahan, O. Influence of activator on the strength and drying shrinkage of alkali-activated slag mortar. *Constr. Build. Mater.* **2009**, *23*, 548–555. [CrossRef]
23. Rashad, A.M. A comprehensive overview about the influence of different additives on the properties of alkali-activated slag-A guide for civil engineer. *Constr. Build. Mater.* **2013**, *47*, 29–55. [CrossRef]
24. Rashad, A.M. A comprehensive overview about the influence of different admixtures and additives on the properties of alkali-activated fly ash. *Mater. Des.* **2014**, *53*, 1005–1025. [CrossRef]
25. Qian, H.; Umar, M.; Khan, M.N.A.; Shi, Y.; Manan, A.; Raza, A.; Chen, G. A State-of-the-Art Review on Shape Memory Alloys (SMA) in Concrete: Mechanical Properties, Self-Healing Capabilities, and Hybrid Composite Fabrication. *Mater. Today Commun.* **2024**, *40*, 109738. [CrossRef]
26. Emdadi, A.; Mehdipour, I.; Libre, N.A.; Shekarchi, M. Optimized workability and mechanical properties of FRCM by using fiber factor approach: Theoretical and experimental study. *Mater. Struct.* **2015**, *48*, 149–1161. [CrossRef]
27. Mehdipour, I.; Vahdani, M.; Libre, N.A.; Shekarchi, M. Relationship between workability and mechanical properties of fibre-reinforced self-consolidating mortar. *Mag. Concr. Res.* **2013**, *65*, 1011–1022. [CrossRef]
28. Si, W.; Cao, M.; Li, L. Establishment of fiber factor for rheological and mechanical performance of polyvinyl alcohol (PVA) fiber reinforced mortar. *Constr. Build. Mater.* **2020**, *265*, 120347. [CrossRef]
29. Ohno, M.; Li, V.C. An integrated design method of engineered geopolymer composite. *Cem. Coner. Compos.* **2018**, *88*, 73–85. [CrossRef]
30. Nematollahi, B.; Qiu, J.; Yang, E.H.; Sanjayan, J. Microscale investigation of fiber-matrix interface properties of strain-hardening geopolymer composite. *Ceram. Int.* **2017**, *43*, 15616–15625. [CrossRef]
31. Shah, I.; Li, J.; Khan, N.; Almujibah, H.R.; Rehman, M.M.; Raza, A.; Peng, Y. Bond-Slip Behavior of Steel Bar and Recycled Steel Fibre-Reinforced Concrete. *J. Renew. Mater.* **2024**, *12*, 167–186. [CrossRef]
32. Liu, Y.; Shi, C.; Zhang, Z.; Li, N.; Shi, D. Mechanical and fracture properties of ultra-high performance geopolymer concrete: Effects of steel fiber and silica fume. *Cemt Concr. Compos.* **2020**, *112*, 103665. [CrossRef]
33. Martinie, L.; Rossi, P.; Roussel, N. Rheology of fiber reinforced cementitious materials: Classification and prediction. *Cem. Concr. Res.* **2010**, *40*, 226–234. [CrossRef]
34. Cui, S.; Fu, F.; Zeng, G.; Chen, Z.; Li, G. Analysis on mechanical property and pore structure of concrete for spraying use in hot-dry environment fiber. *J. Southeast Univ. (Nat. Sci. Ed.)* **2022**, *52*, 43–49.
35. Wang, W.; Huo, J.; Zeng, L.; Liu, W.; Wang, Y. Interlayer Mechanical Characteristics and Anti-penetrability Performance of Combined Shotcrete in Single-shell Lining. *Mater. Rep.* **2023**, *37*, 90–96.
36. Lee, B.Y.; Cho, C.G.; Lim, H.J.; Song, J.K.; Yang, K.H.; Li, V.C. Strain hardening fiber-reinforced alkali-activated mortar—A feasibility study. *Constr. Build. Mater.* **2012**, *37*, 15–20. [CrossRef]
37. Natali, A.; Manzi, S.; Bignozzi, M.C. Novel fiber-reinforced composite materials based on sustainable geopolymer matrix. *Procedia Eng.* **2011**, *21*, 1124–1131. [CrossRef]
38. Zhang, Y.; Wei, S.; Li, Z.; Zhou, X.; Eddie; Chau, C. Impact properties of geopolymer based extrudates incorporated with fly ash and PVA short fiber. *Constr. Build. Mater.* **2008**, *22*, 370–383. [CrossRef]
39. Zhang, Y.; Wei, S.; Li, Z.; Zhou, X. Geopolymer extruded composites with incorporated fly ash and polyvinyl alcohol short fiber. *ACI Mater. J.* **2009**, *106*, 3–10.
40. Hu, F.; Xin, X.; Guo, A.; Yu, Z.Y.; Shao, Q.; Sheikh, M.N.; Sun, Z.H. Effect of mix proportion parameters on chloride erosion resistance of fly ash/slag-based engineered geopolymer composites. *J. Clean. Prod.* **2024**, *438*, 140785. [CrossRef]
41. Yuan, C.F.; Raza, A.; Manan, A.; Manan, A.; Ahmad, S.; Chao, W.; Umar, M. Numerical and experimental study of Yellow River sand in engineered cementitious composite. *Proc. Inst. Civil. Eng.-Eng. Sustain.* **2025**, *178*, 3–20. [CrossRef]
42. Akturk, B.; Akca, A.H.; Kizilkanat, A.B. Fracture response of fiber-reinforced sodium carbonate activated slag mortar. *Constr. Build. Mater.* **2020**, *241*, 118128. [CrossRef]
43. Zhou, X.; Zeng, Y.; Chen, P.; Jiao, Z.; Zheng, W. Mechanical properties of basalt and polypropylene fibre-reinforced alkali-activated slag concrete. *Constr. Build. Mater.* **2020**, *269*, 121284. [CrossRef]
44. Abdollahnejad, Z.; Mastali, M.; Falah, M.; Shaad, K.M.; Luukkonen, T.; Illikainen, M. Durability of the reinforced one-part alkali-activated slag mortar with different fibers. *Waste Biomass* **2020**, *12*, 487–501. [CrossRef]
45. Yuan, C.F.; Xu, S.W.; Raza, A.; Wang, C.; Wang, D. Influence and Mechanism of Curing Methods on Mechanical Properties of Manufactured Sand UHPC. *Materials* **2022**, *15*, 6183. [CrossRef] [PubMed]
46. Raza, A.; Memon, B.A.; Oad, M. Effect of Curing Types on Compressive Strength of Recycled Aggregates Concrete. *Quaid-E-Awam Univ. Res. J. Eng.* **2019**, *17*, 7–12. [CrossRef]
47. *GB/T50081-2019*; Standard for Test Methods of Concrete Physical and Mechanical Properties. China Architecture & Building Press: Beijing, China, 2019.
48. Huang, B.T.; Wu, J.; Yu, J.; Dai, J.G.; Leung, C.K.; Li, V.C. Seawater sea-sand engineered/strain-hardening cementitious composites (ECC/SHCC): Assessment and modeling of crack characteristics. *Cem. Concr. Res.* **2021**, *140*, 106292. [CrossRef]

49. Cheng, T.W.; Chiu, J.P. Fire-resistant geopolymer produced by granulated blast furnace slag. *Miner. Eng.* **2003**, *16*, 205–210. [CrossRef]
50. Chindaprasirt, P.; Jaturapitakkul, C.; Chalee, W.; Rattanasak, U. Comparative study on thecharacteristics of fly ash and bottom ash geopolymers. *Waste Manag.* **2009**, *29*, 539–543. [CrossRef]

polymers

MDPI

Article

Experimental Investigation on Post-Fire Mechanical Properties of Glass Fiber-Reinforced Polymer Rebars

Chanachai Thongchom [1], Lili Hu [2,*], Penpichcha Khongpermgoson Sanit-in [3], Denise-Penelope N. Kontoni [4,5], Nitipong Praphaphankul [6], Koravith Tiprak [6] and Suphanut Kongwat [7,8]

1 Department of Civil Engineering, Faculty of Engineering, Thammasat School of Engineering, Thammasat University, Pathumthani 12120, Thailand; tchanach@engr.tu.ac.th
2 State Key Laboratory of Ocean Engineering, School of Naval Architecture, Ocean and Civil Engineering, Shanghai Jiao Tong University, Shanghai 200240, China
3 Department of Civil Engineering, Faculty of Engineering, Kasetsart University Kamphaeng Saen Campus, Nakhon Pathom 73140, Thailand; fengppcs@ku.ac.th
4 Department of Civil Engineering, School of Engineering, University of the Peloponnese, GR-26334 Patras, Greece; kontoni@uop.gr
5 School of Science and Technology, Hellenic Open University, GR-26335 Patras, Greece
6 Department of Civil and Environmental Engineering, Tokyo Institute of Technology, 2-12-1-M1-23 Ookayama, Meguro-ku, Tokyo 152-8552, Japan; praphaphankul.n.aa@m.titech.ac.jp (N.P.); tiprak.k.aa@m.titech.ac.jp (K.T.)
7 Department of Mechanical Engineering, Faculty of Engineering, King Mongkut's University of Technology Thonburi, Bangkok 10140, Thailand; suphanut.kon@kmutt.ac.th
8 Future Automotive Structure Research Group (FASt), Mobility and Vehicle Technology Research Center, King Mongkut's University of Technology Thonburi, Bangkok 10140, Thailand
* Correspondence: lilihu@sjtu.edu.cn

Abstract: Glass fiber-reinforced polymer (GFRP) rebars are commonly used as an alternative to conventional steel reinforcement in a variety of structural applications due to their superior low cost, strength-to-weight ratio, and durability. However, their mechanical properties after exposure to elevated temperatures, particularly in fire-prone environments, remain a significant concern. To address this concern, the present study focuses on investigating the residual tensile behavior, specifically the tensile strength and elastic modulus, of GFRP rebars exposed to high temperatures that are realistically encountered during fire incidents. The temperature range considered in this analysis spans from 100 °C to 400 °C, with a heating rate of 20 °C/min. The fire duration of 1 h is used. This comprehensive analysis is essential for enhancing our understanding of the performance and applicability of GFRP rebars in fire-prone environments. Based on their actual application in the construction industry, five specimens of three different rebar sizes (16, 20, and 25 mm) were examined for the effect of rebar size on tensile behavior after fire exposure. In addition, the effects were investigated of air- and water-cooling methods on residual tensile behavior. The nominal tensile strength, elastic modulus, and ultimate strain of GFRP rebars at ambient temperature are 930 MPa, 50.2 GPa and 1.85%, respectively. The test results indicated that as the temperature increased to 400 °C, the ultimate tensile strength of the GFRP bars decreased by up to 55%, while the ultimate strain increased by up to 44%, regardless of the cooling method. In addition, when rebars of sizes 16–25 mm were subjected to a 400 °C fire treatment, the smaller the rebar, the greater the percentage of ultimate tensile and strain reduction. These findings hold great significance for the utilization of GFRP bars within the construction industry. This study offers valuable insights into the design of fire-resilient structures, emphasizing the importance of considering rebar size and cooling methods due to their impact on the post-fire tensile strength and strain of GFRP rebars.

Keywords: post-fire strength; glass fiber-reinforced polymer (GFRP) bar; tensile strength; tensile behavior; temperature

1. Introduction

Reinforcing steel in concrete is a vital material used to enhance structural strength. In addition to its usability, it is crucial to continuously assess its deterioration to evaluate the structure's usability and lifespan. However, the deterioration of reinforcing steel in concrete is influenced by various factors, including its mechanical and physical properties, as well as exposure to environmental conditions and different types of disasters, such as fire resistance and corrosion caused by environments with high concentrations of acids or solutions.

Corrosion of the reinforcements in reinforced concrete (RC) has long been recognized as a major factor in the deterioration of the performance, especially in environments where chlorides are present, such as those found near the ocean or where deicing salts are used [1]. Consequently, structural damage can appear regularly throughout the structure's service life. This damage can manifest as cracking and spalling of the cover concrete, ultimately affecting the RC element's load-bearing and shear capacities [2]. Fiber-reinforced polymer (FRP) corrosion-resistant rebars have been considered to address these issues [3]. Nowadays, FRP are employed in a diverse range of civil engineering applications, including as rebars, plates, sheets and pultruded profiles. The use of pultruded profiles is confronted with challenges pertaining to shape distortions over time [4,5]. Consequently, the rebar form, owing to its inherent stability and dependability, has attracted substantial interest and is extensively deployed in infrastructures such as bridges, buildings, and industrial structures. Furthermore, glass fiber-reinforced plastic (GFRP) rebars have been promoted for use in civil engineering due to their low cost, high strength-to-weight and stiffness-to-weight ratios, and corrosion resistance [6]. Despite the fact that the usage of GFRP rebars has been codified [7–9] and is extensively recognized within the industry [10–12], these materials exhibit an inclination for degradation under severe environmental conditions [13]. Research and real-world applications have demonstrated this material's effectiveness in harsh environments involving seawater and sea sand concrete, seismicity, ultraviolet radiation, and water vapor condensation [14–16]. Nevertheless, concrete reinforced with GFRP rebars exposed to high temperatures or fire requires additional research and analysis to accurately assess the residual capacity of the structure and to understand its behavior in greater depth.

Some research indicates that the mechanical properties of GFRP rebars degrade at high temperatures [17–23]. For example, the tensile strength and elastic modulus of GFRP rebars tested at 325 °C decreased by 55% and 30%, respectively [19]. At 350 °C, the initial tensile strength of both GFRP and carbon fiber-reinforced plastic (CFRP) rebars decreased by 45 and 35%, respectively [18]. Notably, the size effect of GFRP rebars is important when applied in real-world engineering. However, only limited studies have focused on the behavior of GFRP rebars with various diameters. For example, only 13 mm diameter rebars were mentioned in the study by Bisby et al. [17], 10 mm diameter rebars are mentioned in the study by Hamad et al. [19], and 9.5 and 12.7 mm diameter rebars are mentioned in the study by Wang et al. [18]. Thus, additional major evidence is required on the behavior of the various GFRP rebar sizes and cooling methods after high-temperature exposure.

In addition, the published literature on conventional steel-reinforced concrete has discussed how the procedure of cooling the temperature of the structure affects the structural performance after high-temperature exposure. For example, Lee et al. [24] found that the bond strength of specimens cooled with water was greater than that of specimens cooled naturally. The distractive effect of water cooling was greater than that of air cooling, particularly for longer heating durations [25]. The post-fire compressive strength, compressive strength, and modulus of elasticity of concrete reinforced with CFRP wraps were affected by the cooling technique, with water cooled specimens experiencing a greater reduction in strength than air-cooled specimens [26]. However, that study focused primarily on FRP sheet reinforcement and not rebar reinforcement.

To comprehensively clarify the mechanical properties of GFRP rebars under elevated temperature conditions—a prevalent scenario in structures susceptible to fire—and in consideration of the industry-standard rebar dimensions and cooling techniques, the current investigation examines the residual tensile behavior of GFRP rebars with diameters ranging from 16 to 25 mm. This examination employs both air-cooling and water-cooling methods. In addition, this research probes into the post-fire mechanical properties of GFRP rebars, with an emphasis on failure mode, the stress–strain relationship, tensile strength, and ultimate strain of GFRP rebars. The investigation was conducted using an ASTM D7205/D7205M-06 standard [27] tensile test with different GFRP rebar diameters, cooling methods, and elevated temperatures (100–400 °C). This study evaluated the tensile performance of GFRP rebars embedded in concrete sleeves subjected to high temperatures with varying rebar sizes and cooling techniques using an ASTM D7205/D7205M-06 standard tensile test [27]. The investigated temperature range provided the information necessary for analyses following various fire scenarios. In the evaluation of how various factors affected the tensile behavior of the GFRP rebars, normalization was applied to the tensile test results in addition to the experimental results to observe changes in the mechanical properties after a fire.

2. Experimental Program

This research investigated the post-fire mechanical properties of GFRP rebars, considering rebar size, maximum temperature exposure, and cooling technique. The experiment comprised 27 testing combinations, with a total of 135 specimens. These specimens consisted of three nominal diameters (16, 20, and 25 mm). Each rebar size was subjected to four different maximum temperatures (100, 200, 300, and 400 °C), while some specimens were maintained at ambient temperature (approximately 28 °C) as the control. After heating, the specimens were further divided into two groups to conduct the two different cooling methods of air or water cooling. Then, the specimens were left for 24 h to ensure that all of them reach a steady state at the ambient temperature before conducting the tensile tests.

2.1. Test Specimens

Rebar sizes of 16, 20, and 25 mm were used because they are the most commonly in the construction industry. In terms of the maximum temperature exposure, a literature review [28] reported that once GFRP rebars are exposed to maximum temperatures between 300 °C and 400 °C, their mechanical properties start to degrade. Thus, the mechanical properties after exposure to maximum temperatures of 100, 200, 300, and 400 °C were investigated in the current study and compared to an unexposed specimen. In addition, both air- and water-cooling methods were employed to simulate real-world scenarios involving post-fire reinforcement of GFRP rebars in structures. Specifically, the water-cooling technique represents situations where the structures have been subjected to fire and subsequently extinguished using water from a fire-fighting system. On the other hand, the air-cooling methods simulate portions of the structures that self-extinguished without the use of any external fire-fighting systems.

The 27 testing combinations, with 5 specimens each, are listed in Table 1. The first three combinations namely G16, G20, and G25 have rebar of sizes 16, 20, and 25 mm in diameter, respectively. Specimens in these three combinations were left at ambient temperature (approximately 25 °C), with no added heating or cooling processes performed. On the other hand, the remaining combinations were divided by considering not only the rebar size but also maximum temperature exposure and cooling approach. To clarify, different combinations were exposed to 4 different maximum temperatures (100, 200, 300, and 400 °C) and 2 different cooling techniques (air or water cooling). Therefore, the remaining combinations were composed of $3 \times 4 \times 2 = 24$ combinations amounting to $24 \times 5 = 120$ specimens. For a clear understanding of the testing methods conducted in each combination, the testing combinations were named systematically. "G dd - ttt c" where "dd", "ttt", and "c" indicate the rebar nominal diameter (mm), maximum temperature exposure (°C), as well as whether water cooled (W) or air cooled (A), respectively. For example, "G20-200W" is the group of five replicates for rebars with 20 mm in nominal diameter that were heated up to 200 °C before conducting the tensile tests after 24 h of water cooling.

Table 1. Test specimens, where RT = room temperature.

Test Code	Diameter (mm)	Temperature (°C)	Cooling Method
G16	16	RT	-
G20	20	RT	-
G25	25	RT	-
G16-100A	16	100	Air
G16-200A	16	200	Air
G16-300A	16	300	Air
G16-400A	16	400	Air
G16-100W	16	100	Water
G16-200W	16	200	Water
G16-300W	16	300	Water
G16-400W	16	400	Water
G20-100A	20	100	Air
G20-200A	20	200	Air
G20-300A	20	300	Air
G20-400A	20	400	Air
G20-100W	20	100	Water
G20-200W	20	200	Water
G20-300W	20	300	Water
G20-400W	20	400	Water
G25-100A	25	100	Air
G25-200A	25	200	Air
G25-300A	25	300	Air
G25-400A	25	400	Air
G25-100W	25	100	Water
G25-200W	25	200	Water
G25-300W	25	300	Water
G25-400W	25	400	Water

All the GFRP rebars were composed of continuous Advantex glass fibers inserted in a vinyl ester matrix using pultrusion. According to the ASTM D792 standard [29], the glass fraction was confirmed as 80.5%. The physical and mechanical properties of GFRP rebars at ambient temperature are shown in Table 2.

Table 2. Physical and mechanical properties of GFRP rebars.

Property	Glass Content	Tensile Strength	Tensile Modulus	Ultimate Strain
Value	80.5% (by mass)	930 MPa	50.2 GPa	0.0185

Figure 1a indicates details of a specimen that was set according to ASTM D7205/D7205M-06 [27]. All the rebars in this research were firmly covered by a steel pipe that was 460 mm long and had a diameter of 48 mm that was filled with mortar at both ends to act as interfaces between rebars and the loading head of the loading machine during the tensile tests. Installing the anchors reduced the stress concentration at the interfaces, which is observed frequently in tensile testing of GFRP rebars; hence, premature tensile failures in the GFRP rebars could be prevented to obtain the full tensile capacity of the specimens. In addition, there was a clear length of rebars between anchors of 380 mm that was heated and attached to a strain gauge to investigate the mechanical properties. The illustration of specimens and anchorages are shown in Figure 1b.

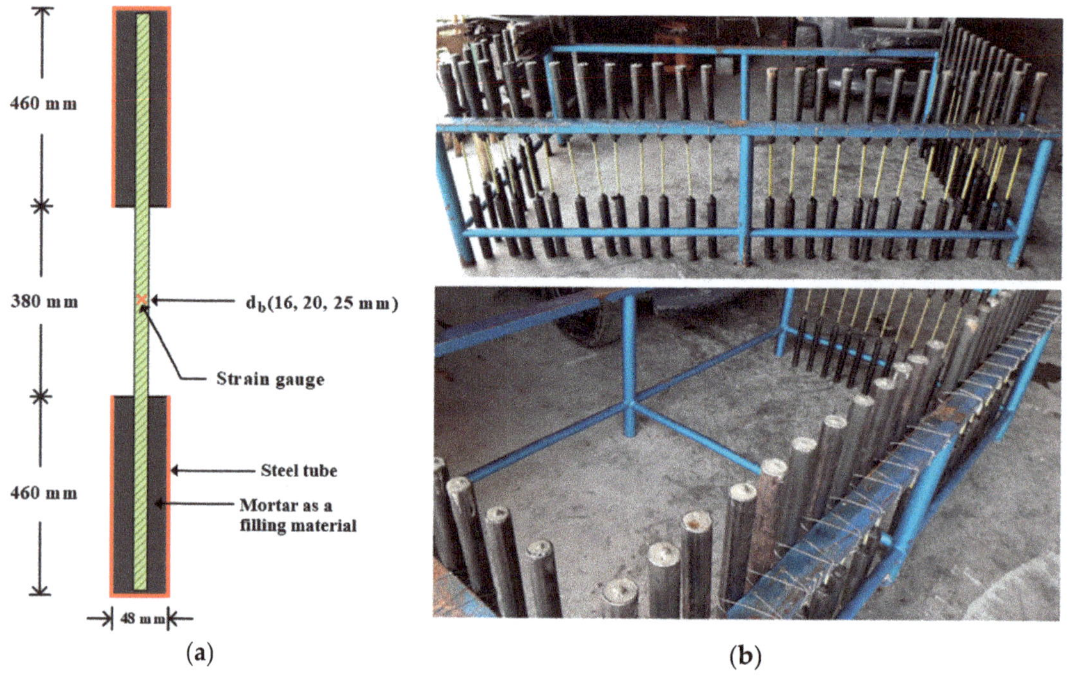

Figure 1. Test specimens: (**a**) details of a specimen (ASTM D7205/D7205M-06 [27]); (**b**) illustration of specimens and anchorages.

2.2. Heating and Cooling

Prior to the experiment, a furnace, capable of heating to 1200 °C, was modified to dimensions of 200 mm, 200 mm, and 200 mm in width, depth, and height, respectively, as shown in Figure 2a to ensure heating occurred only in the desired area. By doing so, the high temperatures were unable to influence and harm the anchors, which played a crucial role during the tensile test as interfaces between the rebars and the loading head of the loading machine.

During the heating process, a GFRP rebar was installed in the modified furnace, as shown in Figure 2b. Then, the specimen was heated to reach the targeted temperature using a heating rate of 20 °C/min [30], as illustrated in Figure 3. After reaching the targeted temperature, the specimen was left for 1 h to simulate a fire at the constant targeted temperature. Then, the designated cooling process was performed on the specimen.

Figure 2. Fire testing: (**a**) internal furnace for heating test; (**b**) fire-testing setup.

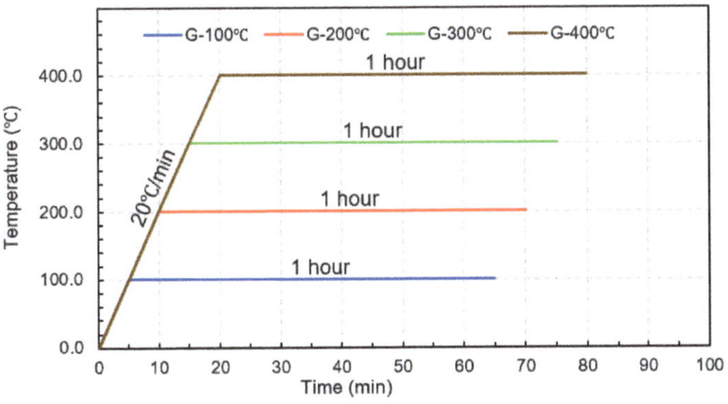

Figure 3. Temperature-time relationship.

Two techniques of cooling were investigated. The first was air cooling, with each specimen left at room temperature of about 25 °C for 24 h after the heating process. The second method used water cooling, where the burnt part of the rebar was submerged in room temperature water immediately after the heating process, and left submerged for 24 h to emulate the influences of cooling on the post-fire mechanical properties of the specimen.

2.3. Tensile Test

After allowing the specimen to cool for 24 h, the standard tensile test according to ASTM D7205/D7205M-06 [27] was conducted, as shown in Figure 4.

First, the anchorage zones of the specimen were firmly attached to the loading head of the loading machine in such a way that slipping during applying the tension force would not occur and the loading direction corresponding to the longitudinal axis of the rebar could be ensured. Next, the burnt rebar surface was polished until a satisfactory level of smoothness was achieved for strain gauge attachment. Finally, before applying the tensile load on the specimen, a strain gauge was attached at the mid-length of the clear range of the rebars for measuring the mechanical properties of the specimen after the fire.

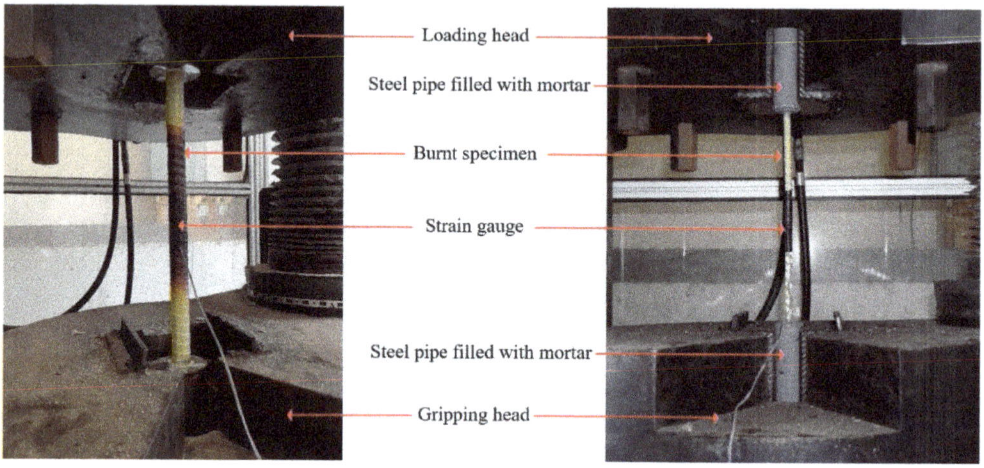

Figure 4. Tensile testing setup.

The tensile test was performed by applying tension force to the specimen through the loading head, which was subsequently increased continuously at a loading rate of 2 mm/min [25] until failure was observed. During the loading process, the tensile stress and tensile strain were recorded to obtain a stress–strain relationship.

3. Results and Discussion

This section presents the results of the tensile tests of the GFRP rebars at elevated temperatures and includes the physical properties and mechanical properties. The physical properties section covers the failure modes of the FRP rebars and changes in the appearance of GFRP rebars under elevated temperatures and after the tensile tests. The mechanical properties section covers the ultimate tensile strength, ultimate strain, and elastic modulus of the GFRP rebars for the evaluated temperature and the normalized mechanical properties. It is noted that, for the purpose of clarity, a normalization approach was employed to facilitate the understanding of the post-fire mechanical properties in relation to the pre-fire properties. The normalization was defined as the ratio of the mechanical properties of the specimen after and before the fire event, was applied to present a clearer picture of the post-fire mechanical properties in comparison to the pre-fire. To illustrate, a value greater than 1.0 in the normalized mechanical properties indicates an improvement in the post-fire properties of the GFRP rebars, whereas a value lower than 1.0 signifies a decrease in the post-fire properties.

3.1. Mode of Failure

After removing the GFRP rebar samples from the heat treatments, samples were observed to have expanded (Figure 5a). This phenomenon may have been attributable to the high values of the resin's coefficient of thermal expansion; this should be noted as the important point since it may contribute to the spalling of concrete under fire conditions. In addition, Figure 5a depicts the GFRP rebars exposed to varying temperatures for the same duration. The rebars exhibited almost no external color change at 100 °C. The surface yellowed when the GFRP rebars were heated to 200 °C and darkened when heated to 300 °C. By the end of their time in the oven at 400 °C, the GFRP rebars had turned an extremely dark color and the surface had turned to ashes in some parts. The hue change was attributable to the decomposition of the polymer matrix during heating. The observed hue shift was clearly correlated with the exposure temperature.

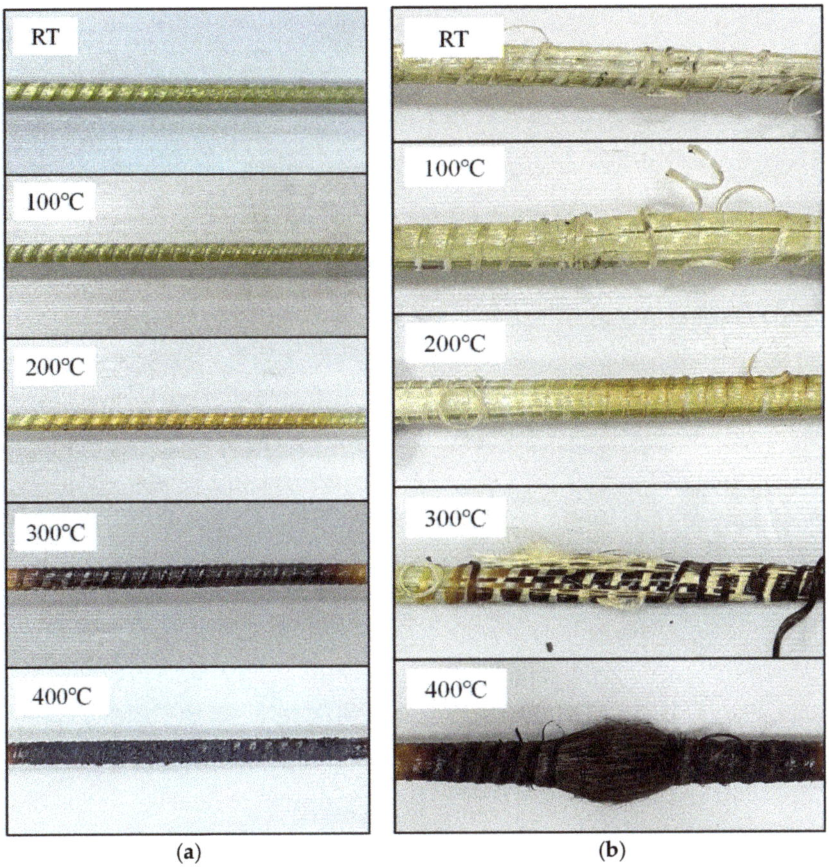

Figure 5. GFRP after fire exposure: (**a**) before tensile test; (**b**) failure mode after tensile test.

In the tensile test, regardless of the elevated temperature or cooling method, all GFRP rebars failed in a similar manner, with fiber rupture at ultimate loads. Due to surface debonding between the fibers and matrix, the failure manifested as a sudden longitudinal central fiber delamination; however, when the GFRP rebars were heated to 300 °C, a similar but more dramatic failure mode was observed, with more fibers collapsing and debonding. At 400 °C, severe delamination was replaced by fiberization. The divergent fibers of each rebar that resulted from the failure are depicted in Figure 5b. Brittle fracture was identified as the predominant mode of failure observed in all the investigated GFRP rebars in this study. This outcome is in accordance with previous studies [31], which have consistently reported the brittle fracture as the common failure mode in GFRP rebars.

3.2. Mechanical Properties

During the tensile test, the stress and strain of specimens were recorded and the elastic modulus was computed. The average (Avg.), standard deviation (S.D.), as well as the coefficient of variation (COV) of the experimental results are listed in Table 3.

Table 3. Experimental results.

Specimen	Tensile Stress (MPa)			Strain (m/m)			Elastic Modulus		
	Avg. (MPa)	S.D. (MPa)	COV	Avg. (m/m)	S.D. (m/m)	COV	Avg. (GPa)	S.D. (GPa)	COV
G16	833.0	9.9	1.2	0.0162	0.00014	0.9	51.5	0.1	0.1
G20	859.4	10.9	1.3	0.0169	0.00016	0.9	50.9	0.2	0.3
G25	955.1	14.1	1.5	0.0184	0.00066	3.6	52.0	2.6	5.1
G16-100A	849.0	11.3	1.3	0.0218	0.00093	4.6	42.1	1.4	3.3
G16-200A	842.0	2.8	0.3	0.0201	0.00030	1.5	40.1	1.4	3.5
G16-300A	809.5	17.7	2.2	0.0232	0.00206	8.9	35.3	3.7	10.5
G16-400A	375.8	17.5	4.7	0.0108	0.00016	1.5	34.7	2.1	6.2
G16-100W	895.4	58.5	6.5	0.0204	0.00075	3.7	43.9	1.2	2.9
G16-200W	899.5	12.0	1.4	0.0230	0.00019	0.8	38.8	0.8	2.2
G16-300W	829.4	15.6	1.9	0.0224	0.00007	0.3	37.0	0.6	1.6
G16-400W	415.7	9.9	2.4	0.0117	0.00083	7.1	35.5	1.7	4.7
G20-100A	898.5	3.7	0.4	0.0198	0.00024	1.2	45.3	0.7	1.6
G20-200A	883.5	7.9	0.9	0.0208	0.00040	1.9	42.4	0.4	1.0
G20-300A	835.7	5.7	0.7	0.0229	0.00140	6.1	36.5	2.0	5.4
G20-400A	500.9	5.4	1.1	0.0135	0.00048	3.5	37.0	0.9	2.4
G20-100W	917.3	10.6	1.2	0.0207	0.00094	4.5	44.3	1.5	3.4
G20-200W	891.3	9.4	1.1	0.0201	0.00003	0.1	44.4	0.5	1.2
G20-300W	849.0	8.3	1.0	0.0212	0.00073	3.3	38.6	0.9	2.3
G20-400W	518.3	9.1	1.8	0.0136	0.00017	1.3	38.2	1.2	3.0
G25-100A	1004.1	11.6	1.2	0.0225	0.00055	2.5	44.8	1.6	3.6
G25-200A	982.8	4.6	0.5	0.0225	0.00111	4.9	43.8	2.4	5.4
G25-300A	947.1	7.5	0.8	0.0233	0.00091	3.9	40.7	1.3	3.1
G25-400A	590.7	9.0	1.5	0.0150	0.00005	0.3	39.3	0.5	1.2
G25-100W	997.6	36.9	3.7	0.0207	0.00056	2.7	48.0	0.5	1.0
G25-200W	989.3	7.5	0.8	0.0211	0.00062	2.9	46.9	1.0	2.2
G25-300W	950.3	22.3	2.3	0.0218	0.00035	1.6	43.5	1.7	1.6
G25-400W	611.3	9.3	1.5	0.0152	0.000001	0.01	40.3	0.6	1.5

Figure 6a–c illustrate the stress–strain curves for G16, G20, and G25, respectively. Based on the test results, the stress–strain relationship of post-fire GFRP bars increased linearly from the beginning until failure. These results agreed well with Spagnuolo et al. [28], who conducted experiments to investigate the residual mechanical behavior of GFRP bars following exposure to temperature treatments ranging from 100 °C to 700 °C. Notably, their study did not specifically explore the effect of the water-cooling technique.

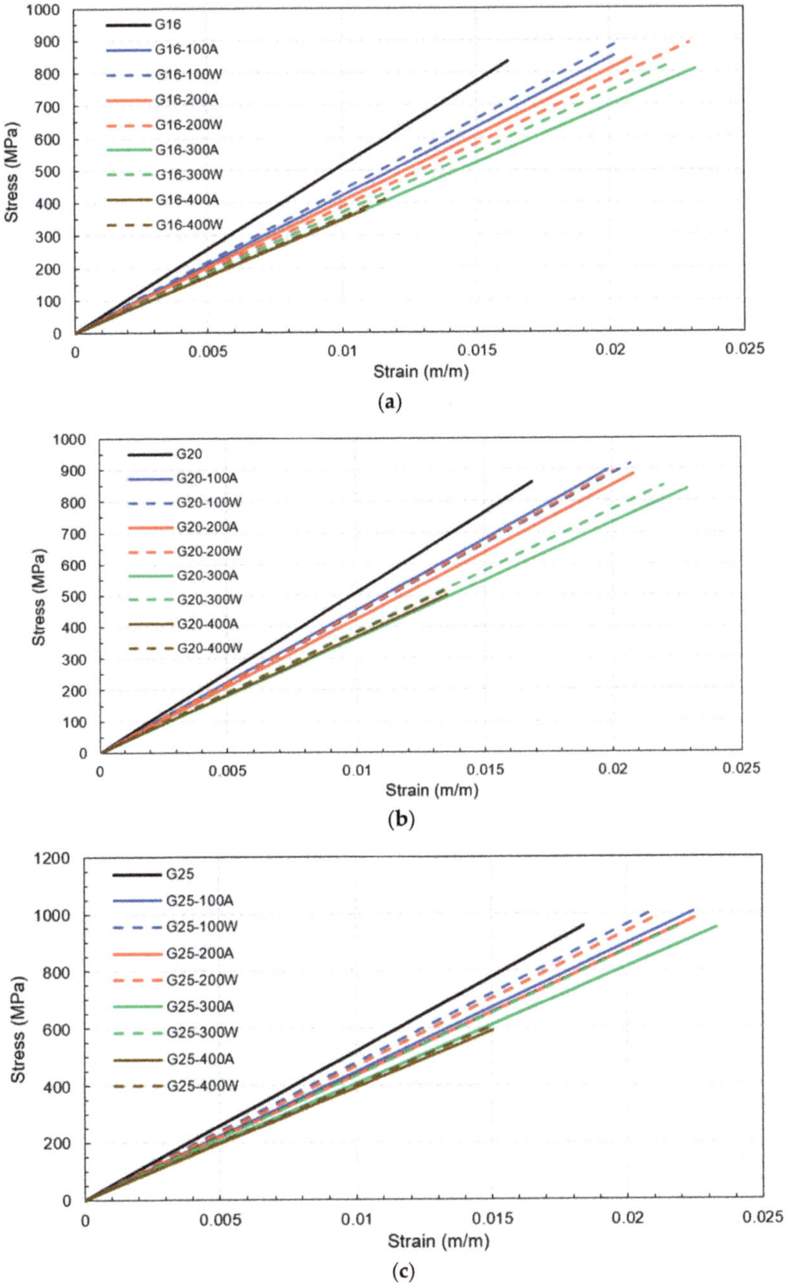

Figure 6. Stress–strain curves for all tested combinations: (**a**) 16 mm (**b**) 20 mm (**c**) 25 mm.

For the maximum temperature exposures of 100 °C, 200 °C, and 300 °C for all rebar sizes and cooling approaches, there was no significant change in the tensile strength compared to the tensile strength of identical-sized rebars before the fire because the maximum temperature exposures did not exceed the critical temperature of between 300 °C and 400 °C; hence, the mechanical properties after being cooled and tested had recovered.

Notably, even though there were slight increments in the tensile strength after exposure to this range of elevated temperatures, these increments might have been inconsistent with the real behavior of the rebars in use because the increases may have been the consequence of better curing of the resin when there was no applied load on the fiber direction of the rebars during the fire, as demonstrated in this study [28]. On the other hand, in cases where the targeted temperature reached 400 °C, the rebars experienced severe damage when they reached the critical temperature range of 300 °C to 400 °C. This resulted in thermal degradation of the polymer and loss of load transfer among the fibers. Consequently, the mechanical properties became non-reversible. This observation is consistent with the findings of a previous study [32] which investigated the tensile, shear, and flexural properties of sand-coated GFRP bars under low temperatures (−100 °C to 0 °C) and elevated temperatures (23 °C to 315 °C). The purpose of that study was to gain insights into the thermal stability of the mechanical properties. Thus, the tensile strengths of G16, G20, and G25 decreased substantially by about 55%, 41%, and 38%, respectively, regardless of the cooling approach, as can be clearly observed from the plot of normalized tensile strength in Figure 7a. Additionally, these numbers implied that the larger the rebar in the range between 16 mm and 25 mm, the smaller the percentage of tensile strength reduction when exposed to 400 °C. It can be seen from Figure 8 that the residual tensile strengths of all rebars after cooling using water were slightly greater than those that were gradually cooled using air.

Considering the ultimate strain, it was clear that all the rebars that were exposed to 100 °C, 200 °C, and 300 °C deformed more than the same-sized rebars before the fire. In other words, rebar ductility increased due to temperature exposure between 100 °C and 300 °C. On the other hand, for maximum temperature exposures to 400 °C, the plot of the normalized ultimate strain in Figure 7b indicates that the ultimate strain of G16, G20, and G25 decreased significantly compared to the pre-fire identical size rebars by about 33%, 20%, and 18%, respectively. Similar to the effects due to the rebar size on the post-fire tensile strength reduction, after exposure to 400 °C, the larger rebars within the range 16–25 mm tended to be more ductile than the smaller rebars in the same range and the percentage of the ductility reduction was relatively lower. In addition, Figure 9 shows that the ultimate strain levels of all rebars after exposure to 400 °C followed by water cooling were more than for the air-cooled rebars, while the ultimate strain levels of the 25 mm rebars after being heated to 100 °C, 200 °C, and 300 °C before being air cooled by air were slightly more than those that were water cooled. However, there was no clear trend observed in the other specimens.

The normalized elastic modulus in Figure 7c demonstrated a clear pattern that as maximum temperature increased, the modulus of elasticity decreased. In addition, some specimens (16 mm and 20 mm rebars heated to 300 °C and 400 °C before being cooled by water as well as all 25 mm water-cooled rebars) had a greater modulus of elasticity compared to those that were air cooled. However, Figure 10 shows that these relationships did not hold true for the 16 mm and 20 mm rebars heated to maximum temperatures of 200 °C and 100 °C, respectively.

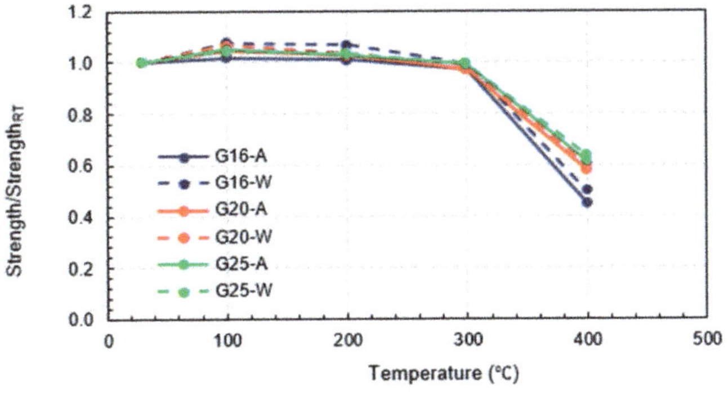

(a)

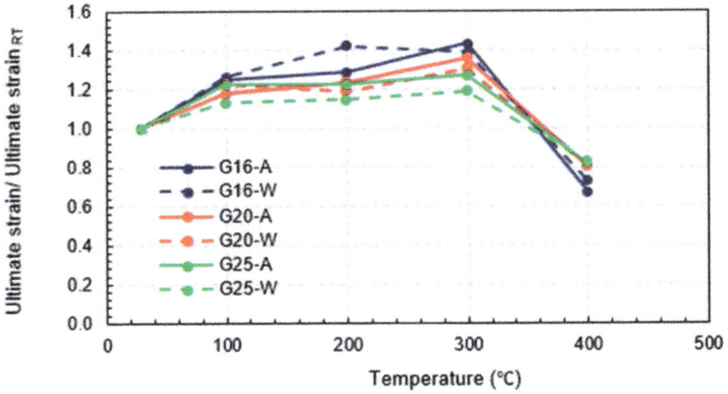

(b)

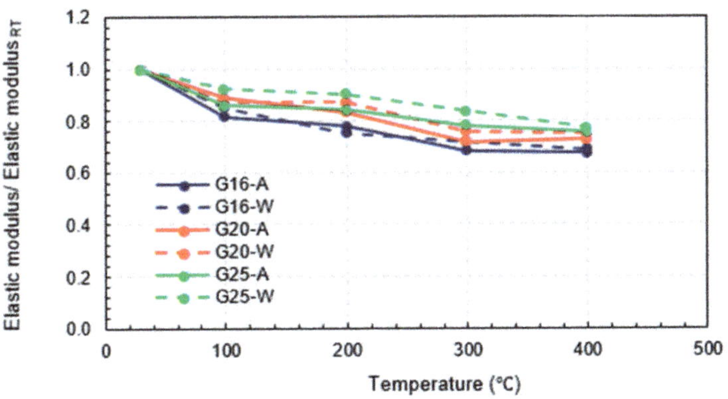

(c)

Figure 7. Results of all testing combinations normalized by results at room temperature: (**a**) normalized tensile strength, (**b**) normalized ultimate strain, and (**c**) normalized elastic modulus.

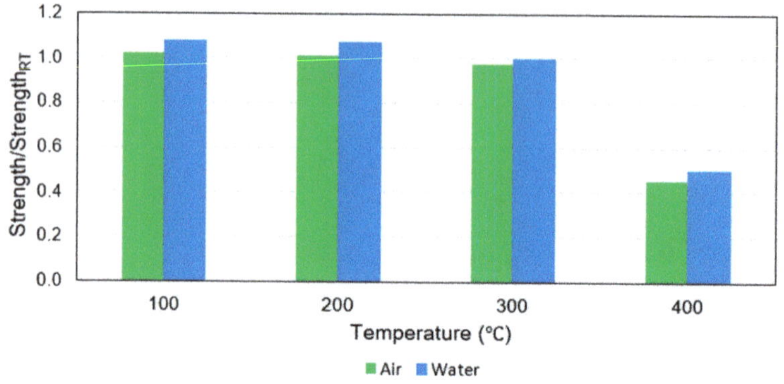

(**a**)

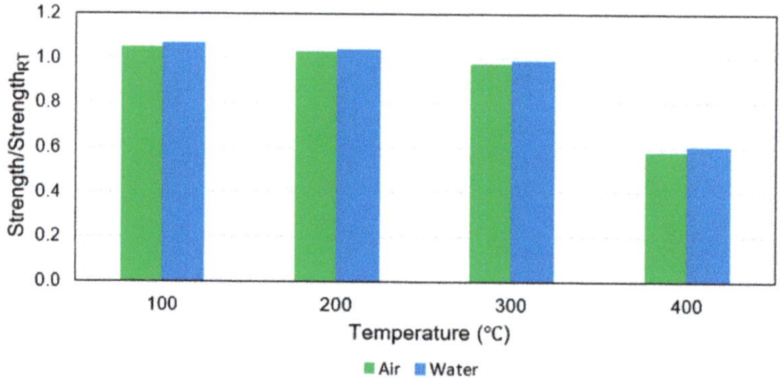

(**b**)

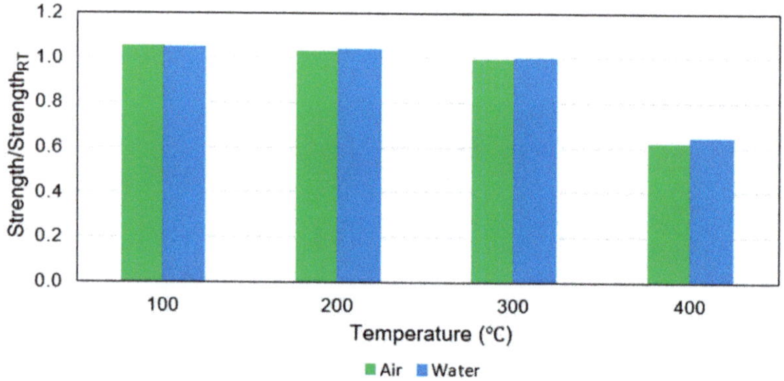

(**c**)

Figure 8. Bar charts of normalized tensile strength: (**a**) 16 mm, (**b**) 20 mm, and (**c**) 25 mm.

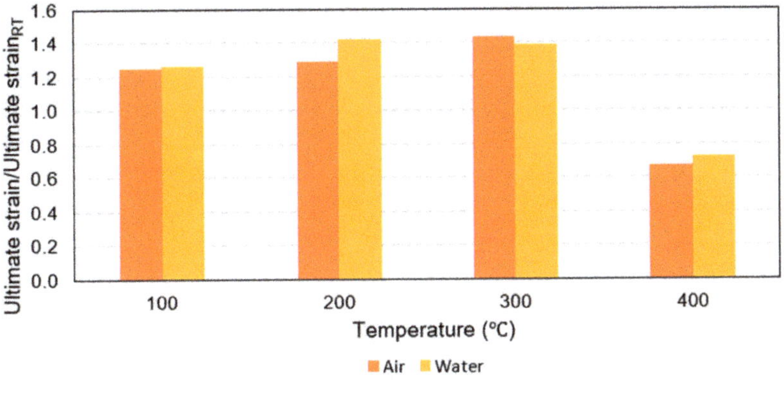

(a)

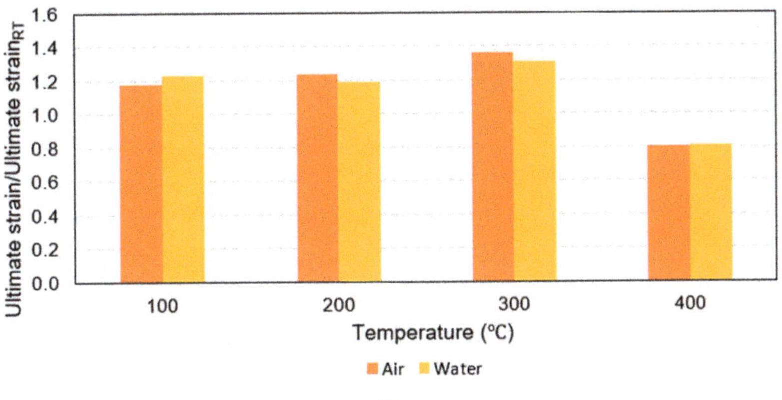

(b)

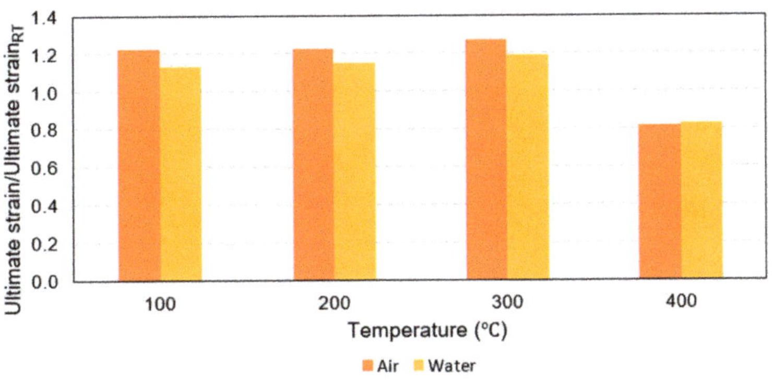

(c)

Figure 9. Bar charts of normalized ultimate strain: (**a**) 16 mm, (**b**) 20 mm, and (**c**) 25 mm.

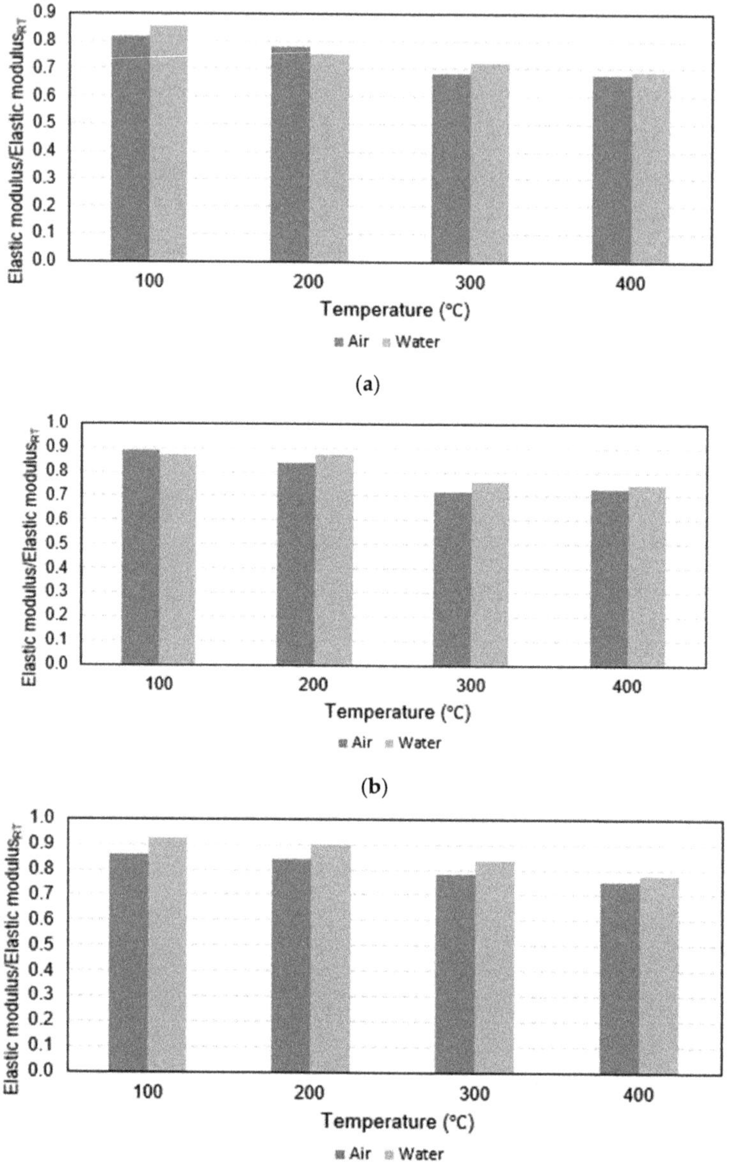

(a)

(b)

(c)

Figure 10. Bar charts of normalized elastic modulus: (**a**) 16 mm, (**b**) 20 mm, and (**c**) 25 mm.

Notably, in the case of traditional steel bars, previous studies such as [33,34] have collected data on the post-fire mechanical properties. It has been observed that the cooling technique exerts a significant influence on these properties. For instance, water-cooled steel bars tend to experience reductions in strength recovery and ductility after exposure to temperatures exceeding 600 °C [33]. However, it is important to clarify that despite a substantial decrease in ultimate strain in water-cooled steel bars by nearly 50%, they still exhibit a notable increase in post-fire strength compared to air-cooled steel bars [34].

However, even though the same results were observed in all the GFRP rebars exposed to more than 300 °C, the differences were far lower than those investigated for steel bars. Thus, it was not reliable to compare the higher strength of the water-cooled GFRP rebars to the air-cooled ones in post-fire evaluation.

3.3. Variation Model of Post-Fire Mechanical Properties

The mechanical properties of the GFRP rebars were normalized using the pre-fire properties to provide an indication of post-fire mechanical properties variation. These normalized properties, namely $N_{i,j}$ where i denotes strength (σ), ultimate strain (ε), and elastic modulus (E) while j denotes the implemented cooling method (A and W for the air and water cooling method, respectively), were plotted against the maximum fire temperature exposures as well as the nominal diameters in Figures 11 and 12, respectively, for air and water cooling, respectively, to show the more apparent course of the correlation between these two parameters and the dependent variables when implementing different cooling methods. After that, the variation trend was fitted with a quadratic response surface model, as shown in Equation (1):

$$N_{i,j}(D,T) = \beta_1 D^2 + \beta_2 T^2 + \beta_3 DT + \beta_4 D + \beta_5 T + \beta_6 \quad (1)$$

where D and T are the nominal diameter (mm) and the maximum fire temperature exposure (°C), respectively. β_1–β_6 are the fitted coefficients of the above quadratic response surface, as listed in Table 4.

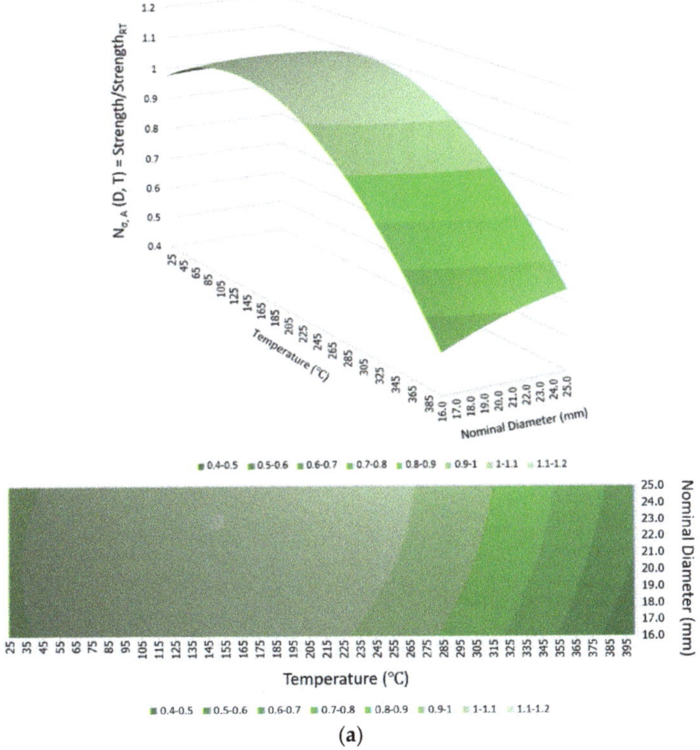

(a)

Figure 11. *Cont.*

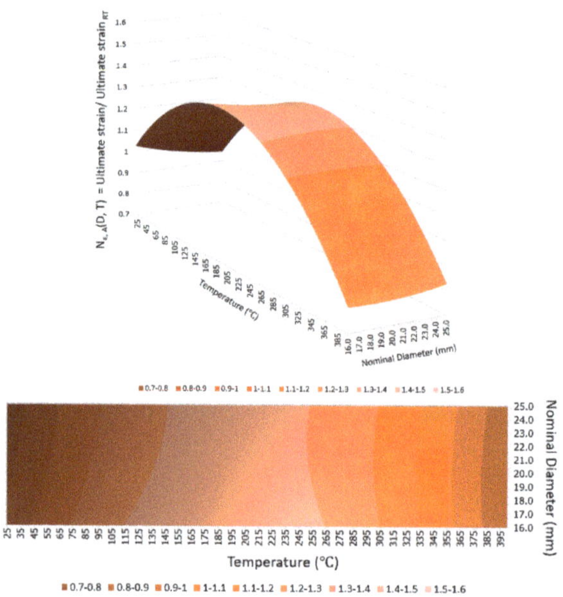

(b)

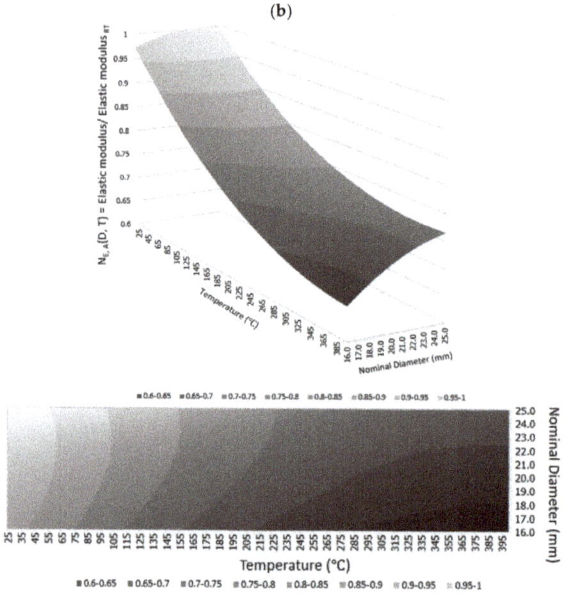

(c)

Figure 11. Models of mechanical properties of various nominal diameter GFRP bars after exposure to different maximum fire temperatures followed by air cooling: (**a**) strength variation, (**b**) ultimate strain variation, and (**c**) elastic modulus variation.

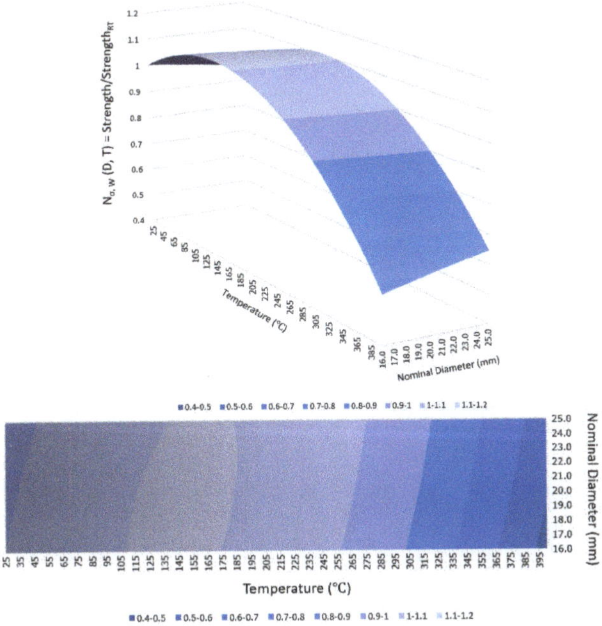

(a)

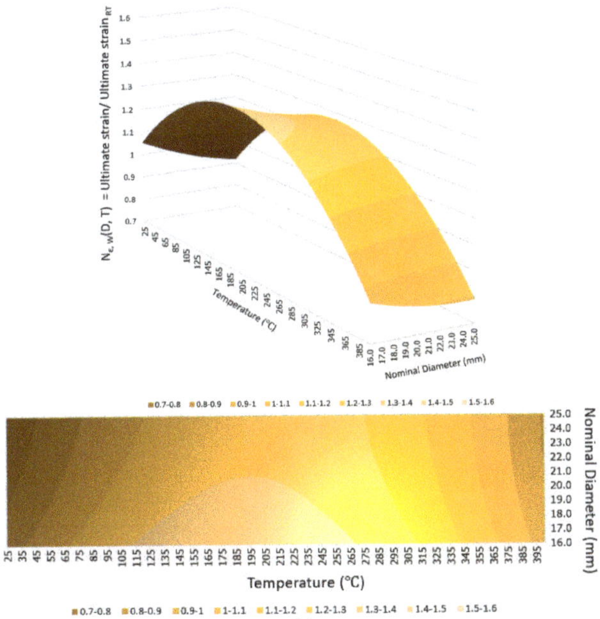

(b)

Figure 12. *Cont.*

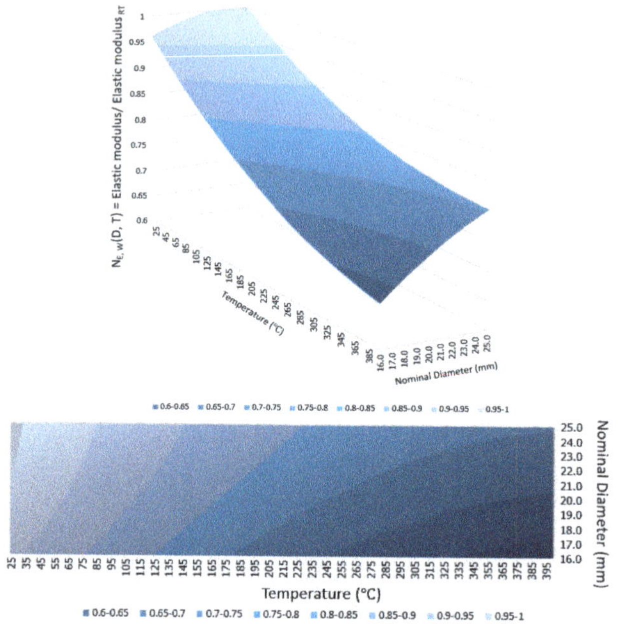

(c)

Figure 12. Models of mechanical properties of various nominal diameters of GFRP bars after exposure to different maximum fire temperatures followed by water cooling: (**a**) strength variation, (**b**) ultimate strain variation, and (**c**) elastic modulus variation.

Table 4. Fitted coefficients of Equation (1), where R^2 = coefficient of determination.

Normalized Mechanical Property	β_1	β_2	β_3	β_4	β_5	β_6	R^2
$N_{\sigma,A}(D,T)$	-7.03×10^{-4}	-7.78×10^{-6}	3.73×10^{-5}	2.65×10^{-2}	1.50×10^{-3}	6.84×10^{-1}	0.93
$N_{\sigma,W}(D,T)$	-1.68×10^{-4}	-8.04×10^{-6}	3.72×10^{-5}	6.98×10^{-4}	1.67×10^{-3}	9.81×10^{-1}	0.94
$N_{\varepsilon,A}(D,T)$	5.75×10^{-4}	-1.28×10^{-5}	3.03×10^{-5}	-3.34×10^{-2}	4.36×10^{-3}	1.30	0.79
$N_{\varepsilon,W}(D,T)$	8.78×10^{-4}	-1.17×10^{-5}	2.07×10^{-5}	-5.14×10^{-2}	4.13×10^{-3}	1.54	0.84
$N_{E,A}(D,T)$	-8.41×10^{-4}	2.00×10^{-6}	2.52×10^{-5}	3.57×10^{-2}	-2.08×10^{-3}	6.58×10^{-1}	0.95
$N_{E,W}(D,T)$	-4.99×10^{-4}	1.33×10^{-6}	2.39×10^{-5}	2.50×10^{-2}	-1.71×10^{-3}	7.23×10^{-1}	0.94

4. Conclusions

This study investigated the post-fire mechanical properties of GFRP rebars, based on rebars with nominal diameters of 16 mm, 20 mm, and 25 mm, maximum temperature exposures between 100 °C and 400 °C, and air-cooling techniques and water-cooling methods. In total, 135 specimens were heated to reach the targeted maximum temperature, followed by continual exposure at that temperature for 1 h before being cooled using the designated cooling approach for 24 h. Finally, tensile tests were conducted on the rebars to obtain the post-fire mechanical properties. The following conclusions were drawn:

The modulus of elasticity of the GFRP rebars decreased as the maximum fire temperature increased compared to the pre-fire modulus of elasticity of the identical-sized rebar.

- Tensile strength reductions were observed in all rebars heated to 400 °C; ductility increased up to 300 °C before decreasing significantly at 400 °C.

- The ductility of GFRP rebars increased due to temperature exposures between 100 °C and 300 °C. However, the ultimate strain decreased significantly after the temperature increased to 400 °C. After exposure to 400 °C, the percentage of tensile strength and ductility reductions decreased as the size of the rebars increased within the range 16–25 mm.
- The post-fire tensile strength of the GFRP rebars that were water cooled was slightly greater than that of the same-sized rebars that were air cooled except for the case of 25 mm bars that were subjected to the maximum temperature of 100 °C. The tensile strength, ultimate strain, and modulus of elasticity of all rebars cooled using water after exposure to temperatures exceeding the critical temperature of between 300 °C and 400 °C, i.e., the case of maximum temperature exposure reach 400 °C in this study were slightly higher than those that had been gradually cooled at ambient temperature. It has been observed that the effects of cooling methods, as observed in this study, are minimal and inconsequential. Furthermore, slight fluctuations in the results were noted.
- The correlations between maximum temperature exposure, rebar nominal diameter, cooling method, and the post-fire mechanical properties were plotted, and variation models were developed by fitting the correlations with the quadratic response surface model. The adoption of this model was based on its capability to provide a reliable reference for design purposes, given its favorable shape.

However, considerations of the physical characteristics, mechanical behavior, and changes in the conditions of reinforcing materials are just some of the important factors that contribute to improving and selecting suitable options for different types of structures. When designing reinforced concrete building elements, it is crucial to consider also the concrete design and to have sufficient concrete cover to ensure proper usage. Incorporating the findings from the study on the behavior of these reinforcing materials into the design process can result in the establishment of appropriate design boundaries. Furthermore, it can lead to a reduction in the consumption of concrete materials by minimizing the concrete cover.

Author Contributions: Conceptualization, C.T. and L.H.; methodology, C.T., P.K.S.-i. and S.K.; validation, C.T., D.-P.N.K., N.P. and K.T.; formal analysis, C.T., N.P. and K.T.; investigation, N.P. and K.T.; writing—original draft preparation, C.T., N.P. and K.T.; writing—review and editing, C.T., L.H., N.P. and K.T.; supervision, L.H., D.-P.N.K., P.K.S.-i. and S.K.; project administration, C.T. and S.K.; funding acquisition, S.K, and L.H. All authors have read and agreed to the published version of the manuscript.

Funding: This research was supported by the Program Fund of Non-Metallic Excellence and Innovation Center for Building Materials (2023TDA5-1) and the Young Elite Scientists Sponsorship Program by CAST (2022QNRC001). This work was also supported by Thammasat University Research Unit in Structural and Foundation Engineering, Faculty of Engineering, Thammasat University, and the Thailand Science Research and Innovation Fundamental Fund fiscal year 2023.

Data Availability Statement: The data presented in this study are available on request from the corresponding author.

Conflicts of Interest: The authors declare no conflict of interest.

References

1. Lu, C.; Yang, J.; Li, H.; Liu, R. Experimental Studies on Chloride Penetration and Steel Corrosion in Cracked Concrete Beams under Drying-Wetting Cycles. *J. Mater. Civil Eng.* **2017**, *29*, 04017114. [CrossRef]
2. Han, C.-G.; Han, M.-C.; Heo, Y.-S. Improvement of residual compressive strength and spalling resistance of high-strength RC columns subjected to fire. *Constr. Build. Mater.* **2009**, *23*, 107–116. [CrossRef]
3. Einde, L.; Zhao, L.; Seible, F. Use of FRP composites in civil structural applications. *Constr. Build. Mater.* **2003**, *17*, 389–403. [CrossRef]
4. Vedernikov, A.N.; Safonov, A.A.; Gusev, S.A.; Carlone, P.; Tucci, F.; Akhatov, I.S. Spring-in experimental evaluation of L-shaped pultruded profiles. *IOP Conf. Ser. Mater. Sci. Eng.* **2020**, *747*, 012013. [CrossRef]

5. Vedernikov, A.; Tucci, F.; Safonov, A.; Carlone, P.; Gusev, S.; Akhatov, I. Investigation on the Shape Distortions of Pultruded Profiles at Different Pulling Speed. *Procedia Manuf.* **2020**, *47*, 1–5. [CrossRef]
6. Debaiky, A.; Cousin, P.; Benmokrane, B. Durability of GFRP bars: A critical review of the literature. *Prog. Struct. Eng. Mater.* **2005**, *7*, 194–209. [CrossRef]
7. *CAN/CSA-S806-02*; Design and Construction of Building Components with Fiber Reinforced Polymers. CSA: Toronto, ON, Canada, 2002.
8. *ACI 440.1R-15*; Guide for the Design and Construction of Concrete Reinforced with FRP Bars. ACI: Dhaka, Bangladesh, 2015.
9. *CNR DT 203-2006*; Guide for the Design and Construction of Concrete Structures Reinforced with Fiber-Reinforced Polymer Bars. CNR: Beijing, China, 2006.
10. Benmokrane, B.; Eisa, M.; El-Gamal, S.; El-Salakawy, E.; Thebeau, D. First use of GFRP bars as reinforcement for continuous reinforced concrete pavement. In Proceedings of the 4th International Conference on FRP Composites in Civil Engineering (CICE2008), Zurich, Switzerland, 22–24 July 2008.
11. Feng, P.; Wang, J.; Wang, Y.; Loughery, D.; Niu, D. Effects of corrosive environments on properties of pultruded GFRP plates. *Compos. Part B Eng.* **2014**, *67*, 427–433. [CrossRef]
12. Alajarmeh, O.; Manalo, A.; Benmokrane, B.; Karunasena, W.; Ferdous, W.; Mendis, P. Hollow concrete columns: Review of structural behavior and new designs using GFRP reinforcement. *Eng. Struct.* **2020**, *203*, 109829. [CrossRef]
13. Bazli, M.; Ashrafi, H.; Oskouei, A.V. Effect of harsh environments on mechanical properties of GFRP pultruded profiles. *Compos. Part B Eng.* **2016**, *99*, 203–215. [CrossRef]
14. Ashrafi, H.; Bazli, M.; Oskouei, A.; Bazli, L. Effect of Sequential Exposure to UV Radiation and Water Vapor Condensation and Extreme Temperatures on the Mechanical Properties of GFRP Bars. *J. Compos. Constr.* **2017**, *22*, 04017047. [CrossRef]
15. Del Zoppo, M.; Di Ludovico, M.; Balsamo, A.; Prota, A.; Manfredi, G. FRP for seismic strengthening of shear controlled RC columns: Experience from earthquakes and experimental analysis. *Compos. Part B Eng.* **2017**, *129*, 47–57. [CrossRef]
16. Wang, Z.; Zhao, X.-L.; Wu, G.; Raman, R.K.; Al-Saadi, S.; Haque, A. Long-term durability of basalt- and glass-fibre reinforced polymer (BFRP/GFRP) bars in seawater and sea sand concrete environment. *Constr. Build. Mater.* **2017**, *139*, 467–489. [CrossRef]
17. Bisby, L.A.; Kodur, V.K.R. Evaluating the fire endurance of concrete slabs reinforced with FRP bars: Considerations for a holistic approach. *Compos. Part B Eng.* **2007**, *38*, 547–558. [CrossRef]
18. Wang, Y.; Wong, P.M.H.; Kodur, V. An experimental study of the mechanical properties of fibre reinforced polymer (FRP) and steel reinforcing bars at elevated temperatures. *Compos. Struct.* **2007**, *80*, 131–140. [CrossRef]
19. Hamad, R.J.A.; Megat Johari, M.A.; Haddad, R.H. Mechanical properties and bond characteristics of different fiber reinforced polymer rebars at elevated temperatures. *Constr. Build. Mater.* **2017**, *142*, 521–535. [CrossRef]
20. Gooranorimi, O.; Claure, G.; De Caso y Basalo, F.; Suaris, W.; Nanni, A. Post-fire behavior of GFRP bars and GFRP-RC slabs. *J. Mater. Civil Eng.* **2018**, *30*, 04017296. [CrossRef]
21. Jarrah, M.; Najafabadi, E.P.; Khaneghahi, M.H.; Oskouei, A.V. The effect of elevated temperatures on the tensile performance of GFRP and CFRP sheets. *Constr. Build. Mater.* **2018**, *190*, 38–52. [CrossRef]
22. Najafabadi, E.P.; Khaneghahi, M.H.; Amiri, H.A.; Estekanchi, H.E.; Ozbakkaloglu, T. Experimental investigation and probabilistic models for residual mechanical properties of GFRP pultruded profiles exposed to elevated temperatures. *Compos. Struct.* **2019**, *211*, 610–629. [CrossRef]
23. Bazli, M.; Abolfazli, M. Mechanical Properties of Fibre Reinforced Polymers under Elevated Temperatures: An Overview. *Polymers* **2020**, *12*, 2600. [CrossRef]
24. Lee, J.; Sheesley, E.; Jing, Y.; Xi, Y.; Willam, K. The effect of heating and cooling on the bond strength between concrete and steel reinforcement bars with and without epoxy coating. *Constr. Build. Mater.* **2018**, *177*, 230–236. [CrossRef]
25. Pul, S.; Atasoy, A.; Senturk, M.; Hajirasouliha, I. Structural performance of reinforced concrete columns subjected to high-temperature and axial loading under different heating-cooling scenarios. *J. Build. Eng.* **2021**, *42*, 102477. [CrossRef]
26. Lenwari, A.; Rungamornrat, J.; Woonprasert, S. Axial Compression Behavior of Fire-Damaged Concrete Cylinders Confined with CFRP Sheets. *J. Compos. Constr.* **2016**, *20*, 04016027. [CrossRef]
27. *Standard D7205*; Standard Test Method for Tensile Properties of Fiber Reinforced Polymer Matrix Composite Rebars. ASTM: West Conshohocken, PA, USA, 2016.
28. Spagnuolo, S.; Meda, A.; Rinaldi, Z.; Nanni, A. Residual behaviour of glass FRP bars subjected to high temperatures. *Compos. Struct.* **2018**, *203*, 886–893. [CrossRef]
29. *Standard D792*; Standard Test Methods for Density and Specific Gravity (Relative Density) of Plastics by Displacement. ASTM: West Conshohocken, PA, USA, 2020.
30. Zhou, F.; Zhang, J.; Song, S.; Yang, D.; Wang, C. Effect of Temperature on Material Properties of Carbon Fiber Reinforced Polymer (CFRP) Tendons: Experiments and Model Assessment. *Materials* **2019**, *12*, 1025. [CrossRef]
31. Al-Salloum, Y.A.; El-Gamal, S.; Almusallam, T.H.; Alsayed, S.H.; Aqel, M. Effect of harsh environmental conditions on the tensile properties of GFRP bars. *Compos. Part B Eng.* **2013**, *45*, 835–844. [CrossRef]
32. Robert, M.; Benmokrane, B. Behavior of GFRP Reinforcing Bars Subjected to Extreme Temperatures. *J. Compos. Constr.* **2010**, *14*(4), 353–360. [CrossRef]

33. Tao, P.Z.; Wang, X.; Uy, B. Stress-Strain Curves of Structural and Reinforcing Steels after Exposure to Elevated Temperatures. *J. Mater. Civil Eng.* **2013**, *25*, 1306–1316. [CrossRef]
34. Lee, J.; Engelhardt, M.; Taleff, E. Mechanical properties of ASTM A992 steel after fire. *Eng. J.* **2012**, *49*, 33–44. [CrossRef]

polymers

MDPI

Article

Research on Hysteretic Behavior of FRP-Confined Concrete Core-Encased Rebar

Jingzhou Lu *, Tong Mou *, Chen Wang, Han Huang and Wenyu Han

School of Civil Engineering, Yantai University, Yantai 264005, China; xyc353934154@163.com (C.W.); hhzgq2580@s.ytu.edu.cn (H.H.); hanwenyu@s.ytu.edu.cn (W.H.)
* Correspondence: jingzhoulu@ytu.edu.cn (J.L.); moutong0035@163.com (T.M.)

Abstract: FRP-confined concrete core-encased rebar (FCCC-R) is a novel composite structure that has recently been proposed to effectively delay the buckling of ordinary rebar and enhance its mechanical properties by utilizing high-strength mortar or concrete and an FRP strip to confine the core. The purpose of this study was to study the hysteretic behavior of FCCC-R specimens under cyclic loading. Different cyclic loading systems were applied to the specimens and the resulting test data were analyzed and compared, in addition to revealing the mechanism of elongation and mechanical properties of the specimens under the different loading systems. Furthermore, finite-element simulation was performed for different FCCC-Rs using the ABAQUS software. The finite-element model was also used for the expansion parameter studies to analyze the effects of different influencing factors, including the different winding layers, winding angles of the GFRP strips, and the rebar-position eccentricity, on the hysteretic properties of FCCC-R. The test result indicates that FCCC-R exhibits superior hysteretic properties in terms of maximum compressive bearing capacity, maximum strain value, fracture stress, and envelope area of the hysteresis loop when compared to ordinary rebar. The hysteretic performance of FCCC-R increases as the slenderness ratio is increased from 10.9 to 24.5 and the constraint diameter is increased from 30 mm to 50 mm, respectively. Under the two cyclic loading systems, the elongation of the FCCC-R specimens is greater than that of ordinary rebar specimens with the same slenderness ratio. For different slenderness ratios, the range of maximum elongation improvement is about 10% to 25%, though there is still a large discrepancy compared to the elongation of ordinary rebar under monotonic tension. Despite the maximum compressive bearing capacity of FCCC-R is improved under cyclic loading, the internal rebars are more prone to buckling. The results of the finite-element simulation are in good agreement with the experimental results. According to the study of expansion parameters, it is found that the hysteretic properties of FCCC-R increase as the number of winding layers (one, three, and five layers) and winding angles (30°, 45°, and 60°) in the GFRP strips increase, while they decrease as the rebar-position eccentricity (0.15, 0.22, and 0.30) increases.

Keywords: FCCC-R; hysteretic behavior; cyclic loading; rebar buckling

1. Introduction

Earthquakes are common natural disasters in nature, and the time and location of their occurrence cannot be accurately predicted by current science and technology. Normally, when a violent earthquake occurs, its tremendous destructive force will cause a devastating blow to the safety of people's lives and property, and reinforced concrete structures are often subjected to enormous cyclic loading forces. This action effect is often characterized by short duration and a small number of cyclic loads, and significant plastic deformation will take place [1,2], causing great damage to the structure in a short time. On the one hand, When the structure is subjected to a low cyclic load after the protective layer of concrete has been peeled away, it is easy for rebar to fracture at an early stage, which will cause the column to lose its lateral bearing capacity and result in structural collapse [3]; however, on the other hand, rebar buckling is also the primary failure mode of reinforced concrete structures under seismic action, which plays an essential role in the seismic resistance of the structure [4–8]. Therefore, improving the fracture properties and buckling resistance of rebar is beneficial to the bearing capacity and stability of building structures under seismic action.

With the aim of ensuring the safety of building structures under seismic action and giving full play to the properties of rebar, numerous methods have been proposed by researchers to enhance the stability of longitudinal rebar. These include densifying stirrups in the plastic-hinge area of the member to limit the buckling of the longitudinally loaded rebar, which is a widely used method at present [9–11]. The appropriate arrangement of transverse rebar significantly improves the seismic performance of reinforced concrete structures but it has always been an issue that makes the binding process of rebar complex and not conducive to the pouring of concrete. In addition, Mitra et al. [12,13] invented an antibuckling structure consisting of a rebar-restraining ring. This research uses a steel pipe to restrain the rebar, which has a fine restraining effect on thinner rebar but is limited to the ordinary reinforced concrete column. FRP has been widely used in construction projects in recent years due to its excellent lightweight, high-strength, corrosion resistance, and ease of cutting properties [14]. Extensive research has been carried out on the application of FRP materials on building structures and reinforced concrete structures [15–19]. As can be seen, FRP materials are widely used in the field of construction engineering due to their unique advantages. Feng et al. [20] inserted a section steel member into an FRP pipe filled with mortar and carried out a monotonic axial compression test. The effects of the cross section of the core steel, slenderness, and FRP fabric layers wrapped at the ends of the specimens were investigated. As a result, these composite components significantly improved the monotonic compressive performance and ductility compared with ordinary section steel members. The monotonic compression ability increased by 44–215% and the ductility increased by up to 877%. Wang et al. [21] proposed a new composite structural FRP-confined concrete core-encased rebar (FCCC-R). Research shows that it can effectively enhance the monotonic compressive capacity of the rebar. The effects of the slenderness ratio, FRP pipe diameter, mortar strength, and rebar-position eccentricity on the monotonic compression capacity of FCCC-R were experimentally studied. An additional parametric study was conducted based on the finite-element model. Lastly, the minimum FRP tube diameter, considering the enhancement effect of monotonic compressive performance, was given, and design curves were proposed for FCCC-R with three types of internal rebar with nominal yield strengths of 400 MPa, 800 MPa, and 1100 MPa. On this basis, Lu et al. [22] also investigated the influence of GFRP strips with different winding layers and angles on the monotonic compression performance of FCCC-R specimens. By performing monotonic compression tests on a series of FCCC-R specimens, it was concluded that their monotonic compression performance increased as the increase in GFRP strips' winding layers and winding angles and decreased as the slenderness ratio increased. A decrease in the slenderness ratio from 22.73 to 15.45 resulted in an increase in the bearing capacity and ductility of the specimens by 33% and 175%, respectively. Wang et al. [23] studied the cyclic axial compressive capacity of RC columns with FCCC-R. It has a much higher axial load capacity compared to ordinary RC columns. A section superposition was proposed to predict load-deformation behavior. Hu et al. [24] investigated the shear behavior of FCCC-R and explored the impact of different influencing factors on its shear performance. Finally, design equations for the shear strength of FCCC-Rs were proposed. It is worth mentioning that FRP composite pipes have excellent corrosion resistance, so they can be applied to marine structures or saline–alkali areas subject to long-term sulfate attacks to improve their durability [25,26]. For example, the strength retention rate of GFRP bars after 100 years of service in a salt solution with an annual mean temperature of 32 °C will still be higher than 82% [27]. The external GFRP strip of FCCC-R can effectively prevent chloride ions and other chemicals from corroding the internal steel bars and the mortar. However, the mechanical response of steel to cyclic loading differs from monotonic loading due to the presence of the

Bauschinger effect, cyclic hardening or softening, and stress relaxation [28–30]. Since the rebar in reinforced concrete structures also bears a cyclic load under the action of seismic force, it should be noted that the mechanical performance response of FCCC-R under a monotonic loading test cannot fully reflect its seismic performance under the action of cyclic seismic force. However, most of the existing research results have focused on the monotonic axial compression and shear properties of FCCC-R, as well as the performance of RC columns with FCCC-R. There is almost no research on the hysteretic performance of FCCC-R under cyclic forces. Rebar plays a crucial role in the seismic performance of RC structures. It is necessary to further understand the changes in the seismic performance of rebar under the constraints of mortar and FRP pipes.

In this study, two cyclic loading systems were used for experimental research on FCCC-R with different slenderness ratios and constrained diameters. Further understanding of the impact of external constraint components on the load-bearing capacity, ductility, and buckling of rebar was obtained through experimental results. In addition, an FCCC-R finite-element model was established using ABAQUS software to predict the stress–strain relationship under cyclic loading. An extended parameter study was also conducted based on the established finite-element model, analyzing the effects of the number of winding layers, winding angle of FRP strips, and rebar-position eccentricity on the hysteretic performance of FCCC-R. Both the experimental results and the proposed finite-element model in this paper have certain reference values for scholars and practitioners.

2. Materials and Methods

2.1. Specimen Design

The steel bars in the FRP-constrained concrete cored rebar in this study were all HRB400 bars of 22 mm diameter. The externally wound FRP strip was a 2 mm thick glass-fiber-reinforced composite (GFRP), and the fiber direction of the GFRP strip was wrapped at an angle of 90° to the radial direction of the specimen. The FCCC-R specimens were formed by inserting rebar into the middle of a GFRP pipe and filling a mortar of C40 strength between the two. Each end of the rebar was set at a length of 100 mm, the outermost 80 mm of which corresponded to the clamp length of the specimen, and the inner 20 mm was the length reserved to ensure that the load was only applied to the rebar. The remaining middle lengths were restrained lengths wrapped by the mortar and the GFRP strip. The configuration of the FCCC-R specimen is shown in Figure 1.

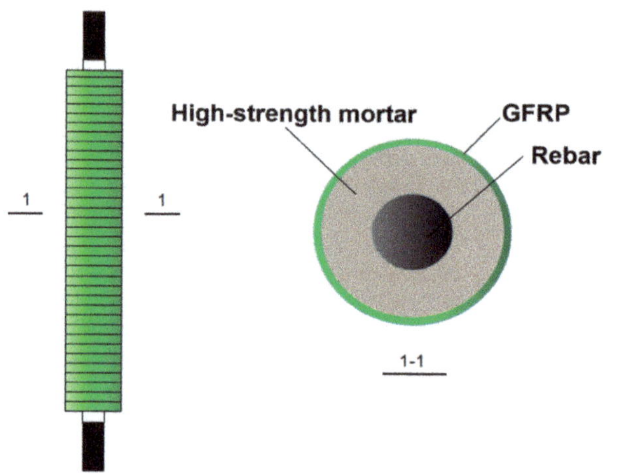

Figure 1. Configuration of FCCC-R specimens.

2.2. Specimen Preparation

The fabrication of FCCC-R specimens in this study primarily included mold fabrication, mortar pouring, mold removal, maintenance, and wrapping of the GFRP strip. The specific production steps are as follows.

Mold making: the fabrication process used PVC pipe as a mold for pouring mortar. Based on the different restraint diameters and restraint lengths of the FCCC-R specimens, the PVC pipes with different diameters were cut to different lengths. Since the ends of the FCCC-R specimens required an exposed steel length of 100 mm, a PVC pipe of that length was intercepted as the pedestal base of the lower end of the mold. One end of the rebar was wrapped with soft plastic and placed in the base to ensure that it was in the center position, as well as to prevent the mortar from permeating in the base. In addition, to keep the mortar from spilling out, the bottom base and the mold were firmly bonded with tape.

Mortar filling, remolding, and maintenance: in order to ensure that the rebar was finally in the center during the pouring process, the upper rebar was fixed with a fixed mold and then the mortar was evenly distributed throughout the mold by using the percussion vibrating method. The specimens were shaped after standing for 24 h and then transferred to the maintenance room for 28 days.

Wrapping of GFRP strips: before wrapping the GFRP strips, the exterior surface of the mortar was wiped with alcohol and the surface was sanded to make it smooth. The fiber strips were then cut according to the different constraint sizes of the FCCC-R specimens. Subsequently, the strips were adhered using a binder, then wrapped at an angle of 90° to the radial direction of the specimen in both fiber directions. The binder was a mixture of epoxy resin and hardener at a ratio of 2:1. Each FCCC-R specimen was glued with 4 layers of fiber strips with a total thickness of 2.8 mm.

2.3. Test Materials

In this study, the FRP-constrained concrete cored rebar specimens were all HRB400 steel bars with 22 mm diameter. The basic mechanical properties were measured in accordance with GB/T228.1-2010, Tensile Test of Metal Materials at Room Temperature. The results meet the requirements of GB50011-2016, Seismic Design Code for Buildings. The mortar material was commercially obtained from Qingdao Zhuonengda Construction Technology Co., Ltd. (Qingdao, China). To facilitate the comparison of the elongation of specimens under different loading systems, the results of the rebar material-properties test are listed in Section 3.4. To determine the basic mechanical properties of the mortar used for this study, three sets of cubic specimens were made according to GB/T50081-2019, Standard for Test Methods for Physical and Mechanical Properties of Concrete, which were tested according to the standard test methods after curing. The average strength of mortar-cube specimens was 43.6 MPa. GFRP was made of unidirectional glass-fiber cloth and epoxy-resin glue produced by Haining Anjie Co., Ltd. (Haining, China)The mechanical properties of the GFRP strips were measured in accordance with the test methods specified in GB/T3354-2014, Test Method for Tensile Properties of Oriented Fiber-Reinforced Polymer Matrix Composites; and GB/T3856-2005, Longitudinal and Transverse Shear Test Method for Polymer Matrix Composites for Tensile Mechanical Properties, Longitudinal and Horizontal Shear Tests, respectively. The basic mechanical properties of the GFRP measured by the test are listed in Table 1.

Table 1. Basic mechanical properties of GFRP strips.

Longitudinal Tensile Strength (MPa)	Transverse Tensile Strength (MPa)	Shear Strength (MPa)	Longitudinal Elastic Modulus (GPa)	Transverse Elastic Modulus (GPa)	Shear Modulus (GPa)	Poisson's Ratio
760.2	55.1	217.1	45.1	2.7	14.9	0.23

2.4. Loading System

The test was conducted using an SDS500 electrohydraulic servo dynamic and static universal testing machine along the radial direction of the specimens for the cyclic loading test; using displacement control loading, the strain rate of the loading rate was 0.00025S^(-1). Ten groups of FCCC-R and ordinary rebar specimens were loaded with constant-amplitude and variable-amplitude mixed loading schemes of two cyclic loading systems, namely the tension–tension and tension–compression equal-amplitude and variable-amplitude cyclic loading systems. The loading systems are shown in Figure 2. Taking the multiple of the yield displacement of the rebar as the unit loading amplitude, the magnitude of each loading amplitude increases in turn, and each one is loaded for two cycles.

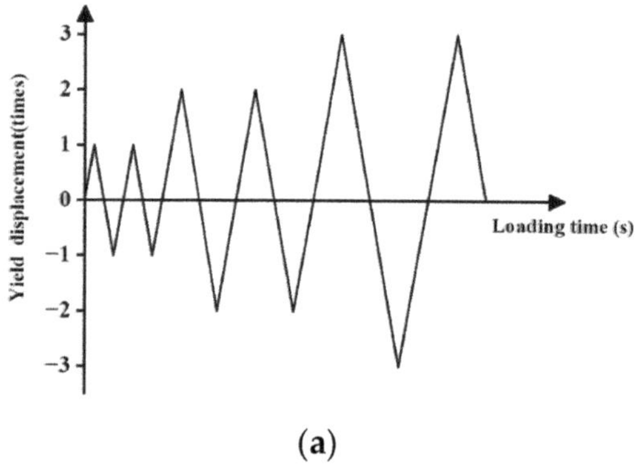

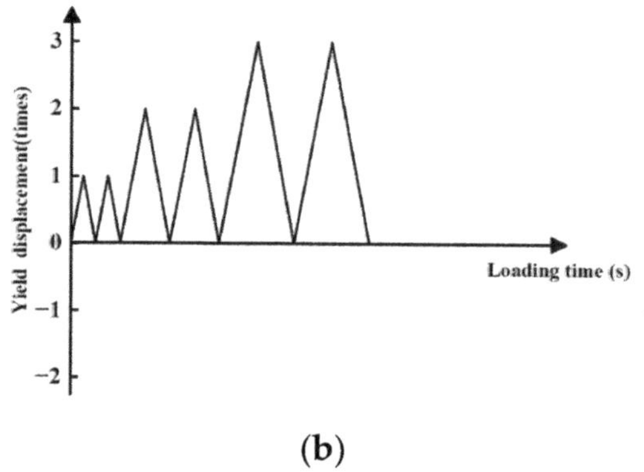

Figure 2. Schematic diagram of cyclic loading systems. (**a**) Tension–compression cyclic loading system; (**b**) Tension–tension cyclic loading system.

The specimens under different cyclic loading systems are listed in Table 2, including 6 groups of FCCC-R specimens and 4 groups of ordinary rebar. Specimens A–F refer to the FCCC-R specimen group, and a–d refer to the ordinary rebar group. The influence of different slenderness ratios of rebar, constraint diameter, and the constraint of composite components on the mechanical properties of rebar under cyclic loading is comprehensively analyzed.

Table 2. Main parameters of cyclic loading specimens.

Specimen Number	Loading Length (mm)	Confined Diameter (mm)	Slenderness Ratio	Unit Yield Displacement (mm)	Loading Speed (mm/min)
A	240	40	10.9	0.54	3.6
B	340	40	15.5	0.765	5.1
C	440	40	20	0.99	6.6
D	540	40	24.5	1.215	8.1
E	440	30	20	0.99	6.6
F	440	50	20	0.99	6.6
a	240	0	10.9	0.54	3.6
b	340	0	15.5	0.765	5.1
c	440	0	20	0.99	6.6
d	540	0	24.5	1.215	8.1

3. Results and Analysis

3.1. FCCC-R Tension and Compression Cyclic Loading Test Phenomenon

The test was loaded according to the predetermined loading system until the specimens were fractured. Due to the different load amplitude of the FCCC-R specimens with different slenderness ratios, their yield displacements also vary. Therefore, the obtained force-displacement curve cannot accurately reflect the hysteresis performance of material unit strain between specimens with different slenderness ratios and is not conducive to their comparison and analysis. It needed to be converted to a stress–strain curve for analysis with the formula $\sigma = F/A$ and $\varepsilon = \Delta l/l$.

Under cyclic loading, the test phenomena of various types of FCCC-R specimens are basically identical. At the initial stage of loading, the deformation of the specimen is mainly elastic. As the loading proceeds, because of the existence of the Bauschinger effect, the specimen soon undergoes plastic deformation during the reverse loading process, and this phenomenon becomes more obvious with an increasing displacement amplitude. The reason for this is that the fiber of the GFRP strip winds circumferentially, and only the bonding force between the fibers produces tension in the radial direction. After the buckling of FCCC-R, the transverse deflection of the specimen increases rapidly and the crack deepens gradually. Finally, with the cracking and destruction of the external restraint components, their utility is greatly reduced and, due to the lack of their support, the damage to the core rebar is aggravated resulting in a large amount of plastic deformation, which causes the specimen to be pulled off. Although the internal filled mortar is crushed and cracked before the fracture of the specimen, the mortar can still maintain a certain shape due to the binding effect of the GFRP strip and it still has a certain binding capacity. It is worth noting that when the rebar is pulled off, there is no "necking" phenomenon as in the monotonous stretching of ordinary rebar. The fracture section of the core rebar of FCCC-R is basically planar. The fracture section of the ordinary rebar is inclined. The fracture morphologies of both are shown in Figure 3. This phenomenon is caused by the "defect" of the material itself. Under repeated loads, some tiny cracks appear at the defect. From that point, the continuous development and deterioration of the microcracks lead to the weakening of the specimen section and the stress concentration at the crack root, and the plastic deformation is limited, leading to the brittle fracture of the rebar after a certain number of cycles.

(a) (b)

Figure 3. Fracture morphology of specimens. (**a**) Fracture morphology of FCCC-R specimen; (**b**) Fracture morphology of ordinary rebar specimen.

3.2. *Tension–Compression Cyclic Loading Test Results*

According to the test results shown in Figure 4, the hysteretic performance of the FCCC-R specimens is significantly better than that of ordinary rebar, which is primarily reflected in the fact that the FCCC-R specimens are able to load more cycles than the ordinary rebar with the same slenderness ratio and has a larger hysteresis envelope area, and the maximum bearing capacity in the compression stage is higher than that of ordinary rebar. At the same time, the hysteretic curve of specimens with a small slenderness ratio and large constraint area is plumper and more symmetrical. The results of this study show that under the same constraint diameter, the maximum compression capacities of the FCCC-R specimens with slenderness ratios of 10.9, 15.5, 20, and 25.5 are 38.7%, 31.1%, 25.0%, and 23.8% higher than those of ordinary rebar, respectively; moreover, the maximum tensile strain before fracture increased by 32.9%, 30.0%, 30.1%, and 20.2%, respectively, and the fracture stress decreased by 42.8%, 1.9%, 11.4%, and 24.8%, respectively. The reduction in the fracture stress of the FCCC-R specimens means that their ductility at fracture is better than that of ordinary rebar [31]. The use of the FRP pipe filled with mortar plays a crucial role in restraining the rebar and fully restrains the premature occurrence of rebar buckling, while also limiting the excessive development of lateral deflections of rebar after buckling. Due to the rapid development of lateral deflections, the root of the crack will propagate and intensify the damage to the rebar, which is more likely to cause brittle fractures to it under stress concentration during reverse tensile loading. In addition, the envelope area of the hysteresis loop increases by 77.8%, 60.9%, 50.4%, and 34.2%, respectively. The comparison of the results is shown in Figure 5. The reason for these phenomena is that the damage to ordinary rebar accumulates faster than that of FCCC-R and the plastic deformation develops more rapidly.

Furthermore, in the process of reciprocating loading, the compression peak stress of members with low slenderness ratios decreases more slowly after buckling, and the rebar damage accumulation is relatively slow. On the contrary, the reduction of compressive peak stress is accelerated, and the damage accumulation is more rapid. To sum up, FCCC-R has better seismic energy-consumption capacity than ordinary rebar. The main mechanical property parameters of FCCC-R cyclic loading with the same constraint diameter in this study are listed in Table 3.

Table 3. Tension–compression cyclic loading test results.

Specimen ID	σ_{max} (MPa)	$\sigma_{max,t}$ (MPa)	α	σ_{max}/σ_d	Maximum Strain Value	Hysteresis Envelope Area (MPa)	Fracture Stress (MPa)
A	626.41	500.32	1.38	1.46	0.029	269.08	74.23
B	584.90	451.25	1.21	1.36	0.029	161.83	114.31
C	520.19	441.08	1.25	1.21	0.029	127.49	91.12
D	503.95	446.40	1.23	1.17	0.027	91.20	66.61
E	446.83	426.74	1.07	1.04	0.025	92.03	71.38
F	582.40	425.90	1.40	1.36	0.027	121.85	51.34
a	454.80	440.82	-	1.06	0.022	151.37	129.83
b	443.39	434.92	-	1.03	0.023	100.58	116.58
c	415.00	434.03	-	0.97	0.023	84.79	102.89
d	408.20	431.23	-	0.95	0.022	67.94	88.52

σ_{max}—the maximum compressive bearing stress; $\sigma_{max,t}$—the maximum tensile stress; α—the ratio of the maximum bearing capacity of the FCCC-R specimens to the ordinary rebar with the same slenderness ratio; σ_d—the stress at the yield point of the reinforcement under monotonic tension.

According to the comparison of skeleton curves, with the increase in the slenderness ratio, there is no obvious change on the tensile side of the skeleton curve except for the specimen with a loading length of 240 mm. The influence on the cyclic loading of tension–compression is mainly reflected in the compression performance. There is a more obvious increase in peak compressive stress. This is because the mortar and GFRP strips have a better restraint effect when the lateral deflection of the rebar occurs, thus, inhibiting the development of the transverse deflection. Moreover, when the rebar does not generate a large transverse deformation, the compression results in an increase in the cross-sectional area, and the friction resistance between the rebar and mortar also increases the compression bearing capacity to a certain extent [30]. When the specimen is in tension, it is more the friction between the mortar and rebar that acts on the restraint of the rebar. The specimen with a loading length of 240 mm is different from other specimens in the result of significantly increasing the maximum bearing capacity at the tensile side. In this case, it is considered that its buckling occurs late, which clearly demonstrates the cyclic strengthening phenomenon, increasing the peak stress of the test piece with the amplitude of each stage of cyclic loading before buckling.

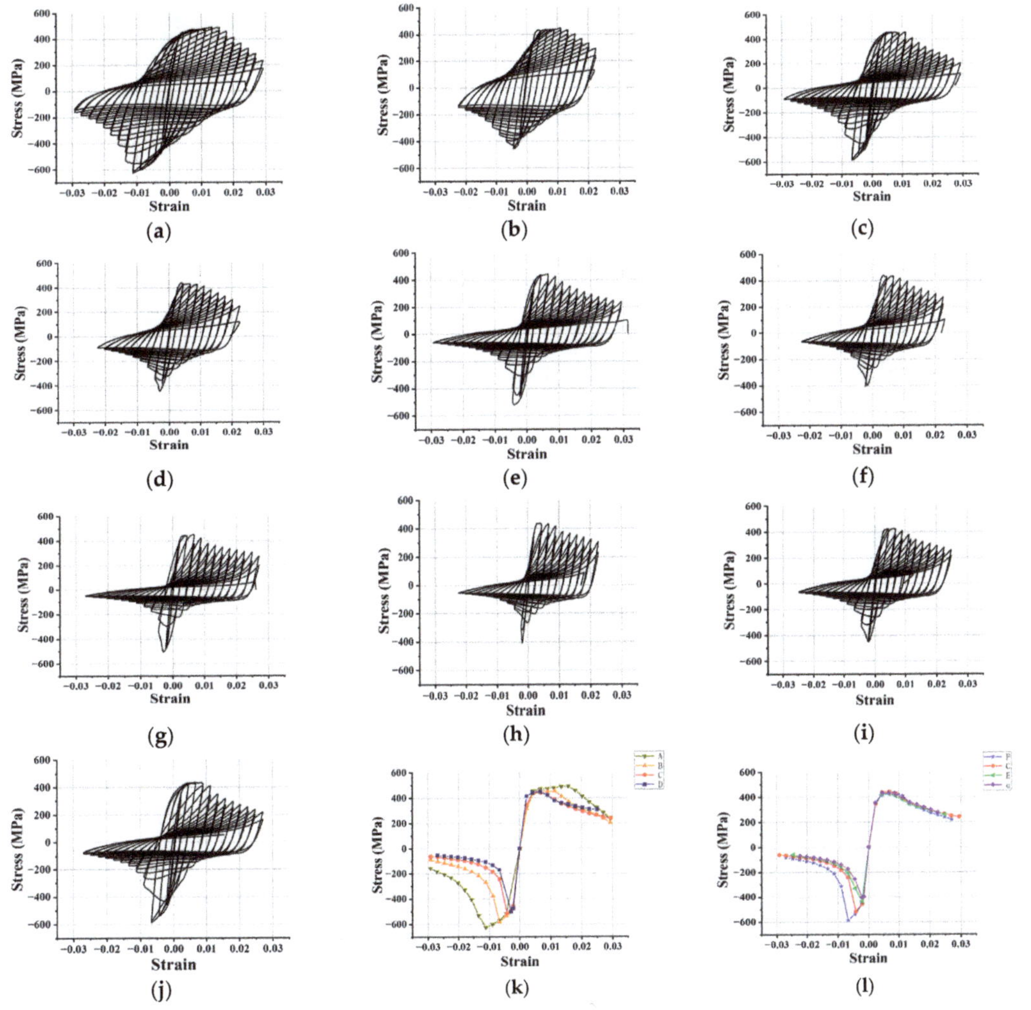

Figure 4. Hysteresis curves of tension–compression cyclic loading. (**a**) Specimen A; (**b**) Specimen a; (**c**) Specimen B; (**d**) Specimen b; (**e**) Specimen C; (**f**) Specimen c; (**g**) Specimen D; (**h**) Specimen d; (**i**) Specimen E; (**j**) Specimen F; (**k**) Comparison of skeleton curves of specimens with different slenderness ratios; (**l**) Comparison of skeleton curves of specimens with different constraint diameters.

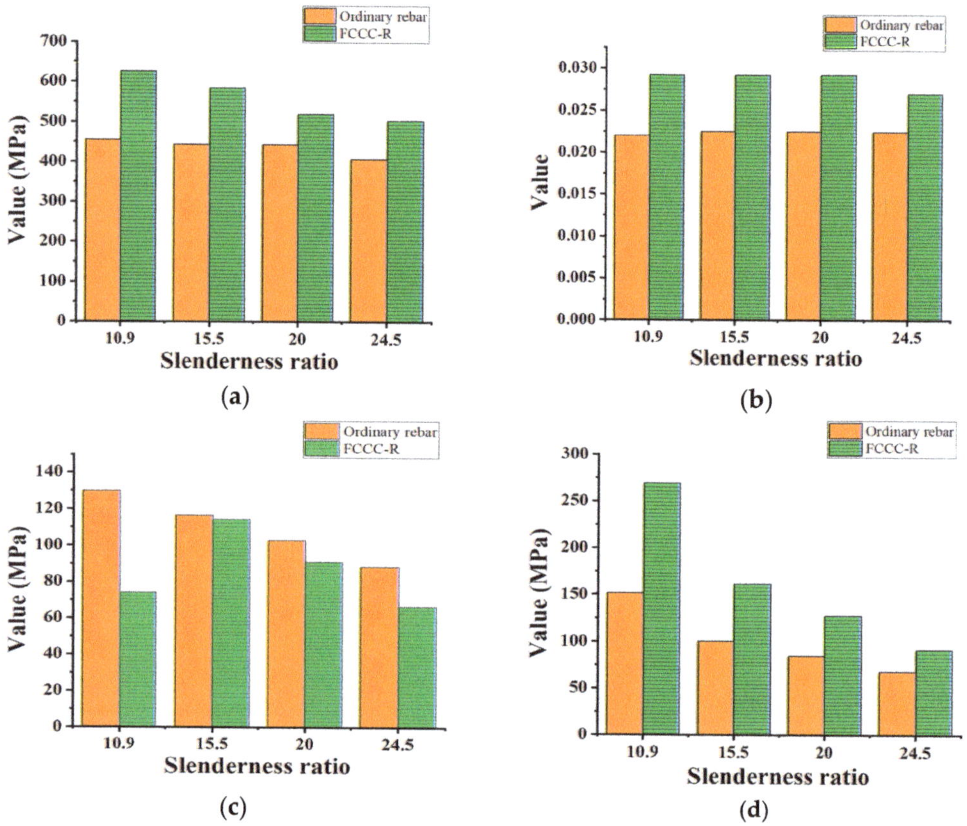

Figure 5. Comparison of basic hysteretic properties between FCCC-R and ordinary rebar under tension–compression cyclic loading. (**a**) Maximum compressive bearing stress; (**b**) Maximum strain before fracture; (**c**) Fracture stress; (**d**) Envelope area of the hysteresis loop.

The influence of the constraint diameter on the hysteretic performance of FCCC-R is also studied here. Three groups of samples with restraint diameters of 30, 40, and 50 mm were used for cyclic loading tests on specimens with a loading length of 440 mm. In agreement with the above results, the constraint diameter had a small impact on the tensile bearing capacity and a large impact on the compressive bearing capacity. With the increase in the restraint diameter, the maximum compressive capacity and the envelope area of the hysteretic ring for each cycle of cyclic loading also increased. The maximum compression bearing capacity of specimens E, C, and F was 7.7%, 30.0%, and 43.7% higher than that of ordinary rebar. The shape of the FCCC-R skeleton curve of each constraint diameter was basically similar, but with the increase in the constraint diameter, the buckling phenomenon appeared later. The FCCC-R specimens with constrained diameters of 30 mm, 40 mm, and 50 mm buckled in the first cyclic compression stage of the second, third, and fourth yield displacement, and the final maximum strain before fracture was 0.025, 0.029, and 0.027. The results are listed in Table 3. In terms of the envelope area of the hysteresis loop and the maximum strain value, the improvement effect of specimen F was not as good as that of specimen C, which does not reflect the regular pattern of the hysteretic performance of FCCC-R increasing with the constraint diameter. Through analysis of its causes, this phenomenon is thought to be caused by the insufficient vibration of mortar in the process of making the specimen or the defects of the rebar itself, which is a premature fracture of the specimen in the process of loading. The premature fracture of specimen F results in a reduction in the number of loading cycles. However, under the same number of loading cycles, the area of the hysteresis loop of specimen F is larger than that of test piece C. Therefore, the FCCC-R specimens with a larger constraint diameter still have better hysteresis performance. When the constraint diameter is 30 mm, the difference between the skeleton curve and the ordinary rebar is small, indicating that the smaller constraint diameter has a very limited effect on improving the hysteretic performance of the rebar. It is, thus, concluded that with the increasing rebar slenderness ratio, only by using the appropriate restraint diameter can an FCCC-R specimen with fine hysteretic performance be obtained. The value relationship of the best constraint diameter under a certain slenderness ratio of rebar needs to be further explored.

3.3. Tension–Tension Cyclic Loading Test Results

The test phenomenon of FCCC-R specimens under tension–tension cyclic loading is roughly the same as that under tension–compression cyclic loading, which is not described in this section. However, the location of the rebar fracture occurs at the end of the restraint length. The fracture position of the ordinary rebar is still in the middle of the specimen. Six groups of specimens with slenderness ratios of 10.9, 15.5, and 20, including ordinary rebar and FCCC-R specimens with a constraint diameter of 40 mm, were loaded under tension–tension cyclic loading until their fracture. The test results are shown in Figure 6.

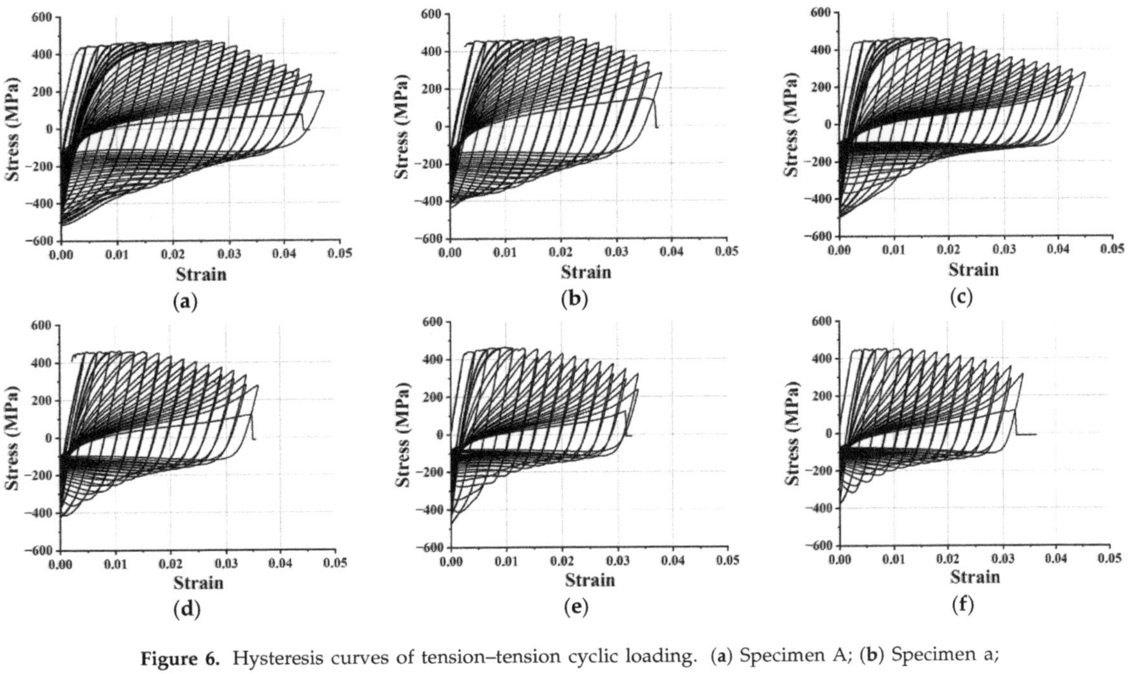

Figure 6. Hysteresis curves of tension–tension cyclic loading. (**a**) Specimen A; (**b**) Specimen a; (**c**) Specimen B; (**d**) Specimen b; (**e**) Specimen C; (**f**) Specimen c.

The hysteretic performance of FCCC-R specimens is significantly improved compared with that of ordinary rebar, which is the same as under tension–compression cyclic loading. Furthermore, their hysteretic properties also increase with the decrease in slenderness ratio. As the slenderness ratio increases under the same constraint conditions, the envelope area of the hysteresis loop of the FCCC-R specimens is 52.0%, 57.1%, and 12.5% higher than that of the ordinary rebar; the maximum tensile strain before fracture increases by 23.7%, 25.0%, and 0% and the maximum compression bearing capacity increases by 19.0%, 18.2%, and 28.3%. The comparison of the results is shown in Figure 7. Moreover, the maximum compressive bearing capacity of the FCCC-R specimens is higher than the yield strength of the ordinary rebar. The restraint components are more conducive to limiting the lateral deflection of FCCC-R with larger slenderness ratios. Among the three, FCCC-R specimens with a slenderness ratio of 20 have a greater improvement in the maximum compressive bearing capacity of the rebar. Under the same slenderness ratio, the maximum tensile bearing capacity of the FCCC-R specimens is still not significantly higher than that of ordinary rebar. Another reason is that the external constraint component plays a very small role in the tensile process. The law is the same as the one illustrated in Section 3.2. Table 4 shows the major hysteretic capacity parameters for both specimen types under cyclic tension–tension loading.

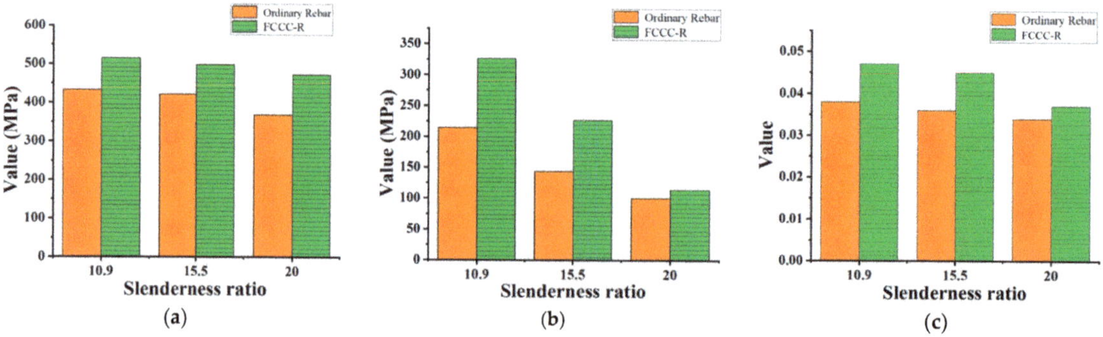

Figure 7. Comparison of basic hysteretic properties between FCCC-R and ordinary rebar under tension–tension cyclic loading. (**a**) Maximum compressive bearing stress; (**b**) Envelope area of hysteresis loop; (**c**) Maximum strain before fracture.

Table 4. Tension–tension cyclic loading test results.

Specimen ID	σ_{max} (MPa)	$\sigma_{max,t}$ (MPa)	σ_{max}/σ_d	Maximum Strain Value	Hysteresis Envelope Area (MPa)	Fracture Stress (MPa)
A	514.6	473.0	1.20	0.047	326.31	71.46
B	498.1	464.5	1.16	0.045	226.80	192.40
C	472.4	453.1	1.10	0.037	114.04	121.07
a	432.5	473.3	1.01	0.038	214.74	130.46
b	421.3	456.2	0.98	0.036	143.72	119.59
c	368.2	448.6	0.86	0.034	100.21	120.74

3.4. Comparison of Elongation under Different Loading Systems

Ductility refers to the ability of the structure or member to resist plastic deformation with no obvious reduction in bearing capacity and refers to the deformation ability of the material, component, and structure after yielding. Elongation is one of the main indicators reflecting the deformation capacity of rebar.

In this study, the monotonic tensile test of ordinary rebar was compared with the elongation of ordinary rebar and some FCCC-R specimens under two cyclic loading systems. The main mechanical properties, and the corresponding elongation of ordinary rebar under monotonic tension, were taken from the test data obtained in the material-property test and directly displayed through the computer-control system connected to the testing machine. The total length of the rebar specimen for the material property test was 500 mm, the loading length was 350 mm, and the loading rate was 0.00025S-1. The test results are listed in Table 5. The specific values are shown in Table 6.

Table 5. Experimental results of monotonic tensile rebar.

Elastic Modulus (MPa)	Yield Strength (MPa)	Tensile Strength (MPa)	Maximum Stress Elongation (%)	Elongation at Break (%)
190.2	429.0	557.5	13.7	22.8

Table 6. Comparison of cyclic loading elongation.

Specimen Number	Elongation at Break (%)	Tensile Ultimate Strength Elongation (%)	Maximum Elongation (%)	Number of Cycles
LL-A	4.3	2.4	4.7	39
LL-B	4.2	1.7	4.5	37
LL-C	3.1	1.1	3.7	28
LL-a	3.7	2.2	3.8	31
LL-b	3.5	1.5	3.6	29
LL-c	3.1	0.8	3.4	27
LY-A	2.4	1.6	2.9	26
LY-B	2.9	1.1	2.9	25
LY-C	3.1	0.6	2.9	26
LY-a	2.2	1.1	2.3	20
LY-b	2.2	0.4	2.3	19
LY-c	2.3	0.6	2.3	20

LL—tension–tension cyclic loading system; LY—tension–compression cyclic loading system.

By comparing the elongation of different specimens under different loading systems, in general, the elongation of rebar under cyclic loading is significantly different from that under monotonic loading; the former is significantly smaller than the latter. Based on the comparison of the maximum elongation of the specimens shown in Figure 8, under two different cycling systems, the elongation of rebar under the tension–tension cyclic loading is greater than that under the tension–compression cycling loading. Compared with tension–compression, under tension–compression cyclic loading, the maximum elongation of FCCC-R specimens with slenderness ratios of 10.9, 15.5, and 20 increased by 62.1%, 55.2%, and 27.6%, respectively. The rebar was enhanced by 65.2%, 56.5%, and 47.8%.

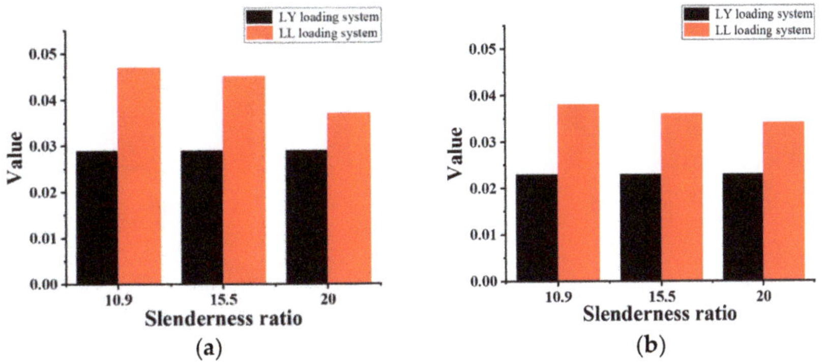

Figure 8. Comparison of maximum elongation of specimens under two loading systems. (**a**) FCCC-R; (**b**) Ordinary rebar.

There is also a difference in the stress–strain relationship of the specimens under the two cyclic loading systems. This is because the elongation and the stress state of the rebar are related to the cyclic loading history of the rebar [27,28], which indicates that the damage accumulation of the tension–compression system is more severe than that of the tension–tension system. However, regardless of the system, the elongation of FCCC-R specimens under the same loading length is greater than that of ordinary rebar. It can be seen that its ductility is better than that of ordinary rebar and the external constraint improves the ductility of the rebar. However, there is still a gap between the elongation of FCCC-R specimens under cyclic loading systems and that of rebar specimens under monotonic tension. The fracture elongation of FCCC-R specimens under tension–tension cyclic loading can reach approximately 20~30% of the elongation at a maximum force under the monotonic loading of ordinary rebar. This value is about 15~20% under tension–compression cyclic loading. In other words, the existence of external restraint components improves the ductility of rebar under cyclic loading but there is still a big gap between the elongation of rebar under monotonic loading.

3.5. Comparison of Mechanical Properties under Different Loading Systems

The objective of this section is to compare and analyze the maximum compressive bearing capacity and envelope area of the hysteresis loop of specimens under two different cyclic loading systems. The results are shown in Figure 9.

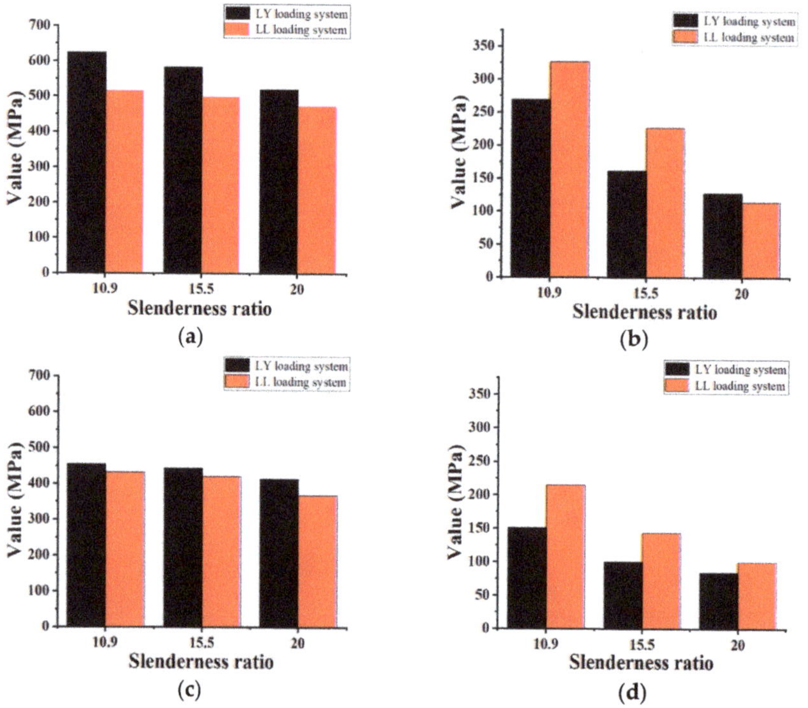

Figure 9. Comparison of mechanical properties of specimens under two loading systems. (**a**) Maximum compressive bearing capacity of FCCC-R; (**b**) Envelope area of the hysteresis loop of FCCC-R; (**c**) Maximum compressive bearing capacity of ordinary rebar; (**d**) Envelope area of the hysteresis loop of ordinary rebar.

Due to the more obvious cyclic hardening effect of steel, the maximum compressive bearing capacity of the specimens under tension–compression cyclic loading is greater than that under tension–tension cyclic loading. The maximum compressive bearing capacity of specimens A–C and a–c increased by 21.7%, 17.4%, 10.1%, 5.1%, 5.2%, and 12.7%, respectively. In contrast, the envelope area of the hysteresis loop was relatively larger under the tension–tension system. Except for specimen C, which decreased by 10.5%, the other five groups of specimens increased by 21.2%, 40.1%, 41.8%, 42.8%, and 18.2%. The reason for this phenomenon is that the tension–tension system has a slower accumulation of damage to the rebar, so more cycles can be loaded.

The basic mechanical properties of specimens C and c under tension–compression cyclic loading are compared with the monotonic compression performance in [22]. From Figure 10, it can be seen that the maximum compressive bearing capacity of specimen C under tension–compression cyclic loading increased by 9.9% compared to monotonic loading, from 473 MPa to 520 MPa. In addition, the rebars are more prone to buckling under cyclic loading. The buckling strain of the FCCC-R is 0.0061 under monotonic loading, while under cyclic loading, this value is only 0.0053. Similarly, the maximum compressive bearing capacity of the ordinary rebar under tension–compression cyclic loading increased from 362.1 MPa under monotonic loading to 415 MPa, and the buckling strain decreased from 0.0019 to 0.0017. From a comparative perspective, whether under monotonic compression or cyclic loading, FCCC-R has a significant effect on improving the maximum compressive bearing capacity of rebars and delaying their buckling. Therefore, FCCC-R has a good effect on fully utilizing the mechanical properties of internal rebars.

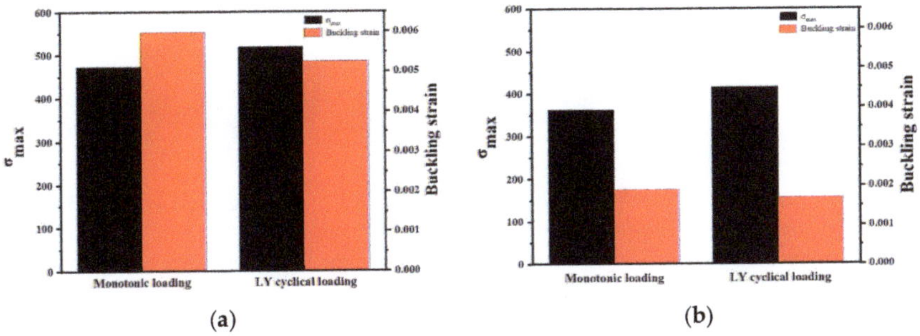

Figure 10. Comparison of mechanical properties between monotonic loading and cyclic loading of FCCC-R. (**a**) FCCC-R; (**b**) Ordinary rebar.

4. Finite-Element Simulation and Parametric Analysis

4.1. Material Constitutive Model

In this study, the finite-element numerical simulation software ABAQUS was used to simulate the hysteretic curve of FCCC-R specimens under tension–compression cyclic loading. We divided the FCCC-R model into three parts—steel, mortar, and a GFRP tube—to establish a finite-element model, in which the mortar and core rebar models were established using solid elements. Due to the low thickness of the GFRP strip compared to the constrained component size, the shell element was used to establish the GFRP tube model. The element type used for the solid element was C3D8R, whereas S4R was used for the shell element. All elements were approximately 6 mm in size. All components were assembled and grids were divided according to the different types of FCCC-R forms. The FCCC-R finite-element model is shown in Figure 11. After observing that the FCCC-R specimen had been damaged after the test, it was found that there was no obvious slippage in the interface between the mortar and the rebar, nor in the interface between the mortar and the GFRP strip. For this reason, in ABAQUS, the "tie" function was used to bind the three parts into pairs. Furthermore, in order to trigger the specimens' instability during cyclic loading, the shape of the first-order buckling mode with a loading length of 1/1000 in the buckling analysis results was introduced into the finite-element model as an initial geometric defect [22].

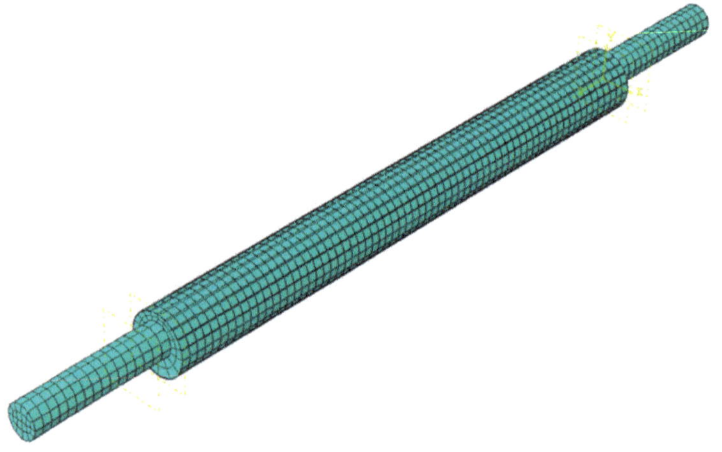

Figure 11. The finite-element model of FCCC-R.

4.2. Material Constitutive Model

The concrete damage plasticity (CDP) model included in the ABAQUS material library was used to define the three-dimensional constitutive relationship of mortar. The CDP model can be used to simulate the mechanical behavior of concrete structures and composite structures under reciprocating loads and is widely used in seismic analysis. Its main parameters, and the compressive and tensile constitutive relationships, are described in [21]. In the experiment, the filling mortar was subjected to low circumferential confinement from the external composite pipe. Therefore, the influence of the plastic-strain law of concrete on the constraint effect was not considered in this simulation. For the GFRP materials, the mechanical behavior of GFRP was divided into elastic and damage destruction stages. Due to the anisotropic elasticity of GFRP, the model used engineering constants to define the tensile modulus of elasticity, shear modulus, and Poisson's ratio in three directions. The Hashin damage criterion has been widely used in progressive damage analysis of composite materials and in defining the failure mode of GFRP strips during the damage stage [32,33]. The main parameters in the Hashin damage criterion were taken based on the material property test results, which are shown in Table 1. The Chaboche [34] steel-plastic constitutive model was used to simulate the stress–strain curve of the rebar under cyclic loading. The model assumes von Mises yield criterion and an associative flow rule is assumed. The characteristics of a material are defined by a combined isotropic/kinematic hardening model. Figure 12 is a diagram of the isotropic/kinematic hardening model. This model can effectively simulate the Bauschinger effect and the cyclic hardening or softening of rebar during cyclic loading. The constitutive model parameters of the rebar under cyclic loading were obtained by fitting experimental data. The specific method is shown in [2,35–37]. In this study, a total of three sets of backstresses were taken. When setting rebar material properties in ABAQUS, the "combine" function was used to set the material strengthening model; then, the fitted constitutive model parameter values in Table 7 were input.

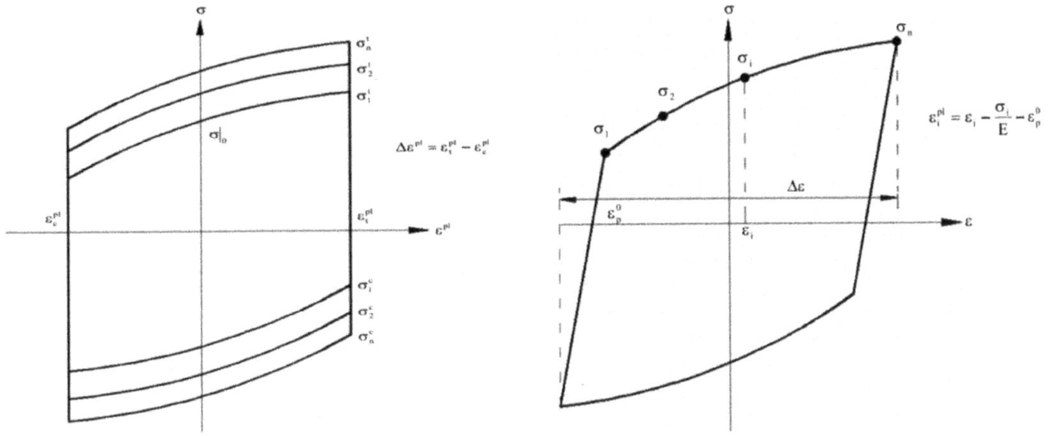

(a) (b)

Figure 12. Rebar finite-element model parameter calibration [2]. (**a**) Isotropic hardening component; (**b**) Kinematic hardening component.

Table 7. Rebar hardening parameter calibration.

Constitutive Model Parameters	$\sigma\|_0$ (MPa)	Q_∞ (N/mm^2)	b_{iso}	$C_{kin,1}$ (N/mm^2)	γ_1	$C_{kin,2}$ (N/mm^2)	γ_2	$C_{kin,3}$ (N/mm^2)	γ_3
Rebar	429	10	1.2	5000	110	6773	116	2854	34

4.3. Comparison of Test and Simulation Results

The objective of this section was to simulate the hysteretic curve of FCCC-R specimens subjected to tension–compression cyclic loading. At the same time, a unit loading amplitude of twice the yield displacement loading system was introduced to conduct a finite-element simulation on specimen C and we labeled this group of samples as specimen G. The simulation results were compared with test data to further verify the accuracy of cyclic strengthening parameters. When the FCCC-R specimens were loaded to the eighth yield displacement, their hysteretic performance was greatly reduced, which also saved model calculation time. Therefore, this study simulates hysteretic curves of the first-eight yield displacements of FCCC-R specimens A–F. For specimen G, it took the same loading time as specimen A–F models for simulation. The simulation and test results of a total of seven groups of specimens were compared, as shown in Figure 13. It can be seen that the simulation results reflect the cyclic hardening and softening, unloading stiffness, peak load, and buckling phenomenon of the rebar under cyclic loading, all of which are in good agreement with the test results. Table 8 summarizes the peak load and hysteresis-loop envelope area obtained by finite-element calculation and compares them with the test results. As can be seen, the discrepancy between the calculated value and the test value is approximately 10%, and only the difference between the calculated value of the hysteresis-loop envelope area of individual specimens and the test value is more than 20%. In general, the simulation and test results agree well, indicating that the constitutive relationship and strengthening parameters used in this study can accurately simulate the stress–strain curve of FCCC-R specimens under tension–compression cyclic loading.

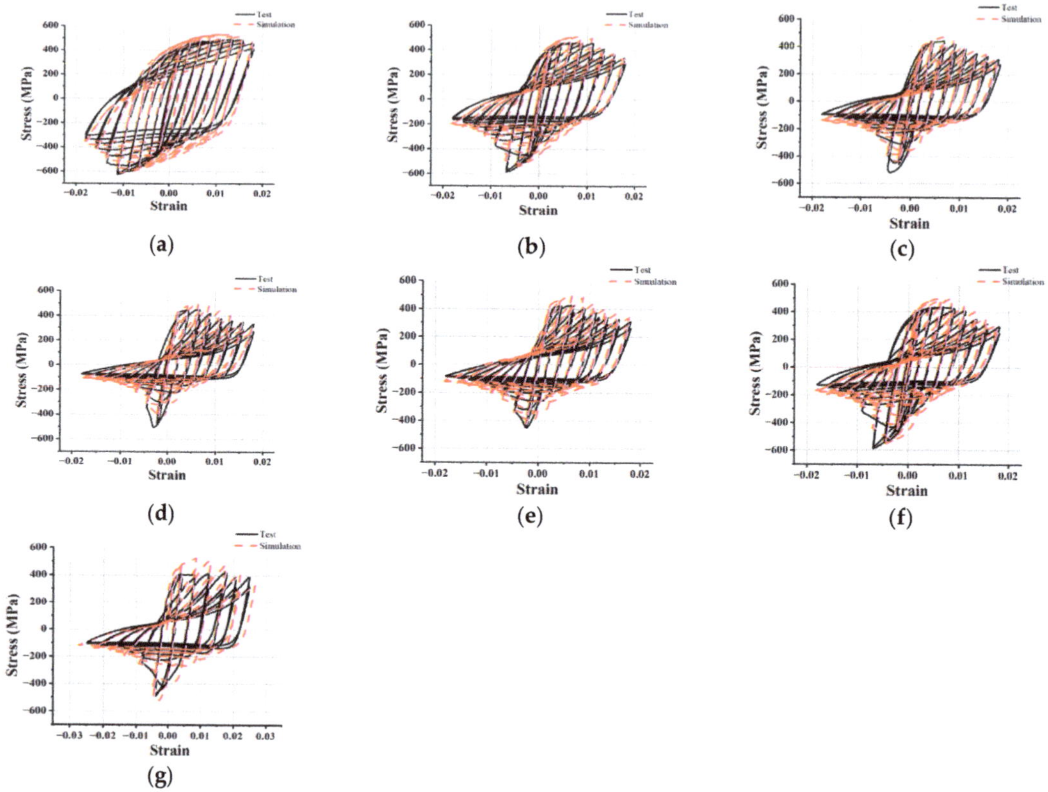

Figure 13. Test and simulation hysteretic curve comparison for FCCC-R specimens. (**a**) Specimen A; (**b**) Specimen B; (**c**) Specimen C; (**d**) Specimen D; (**e**) Specimen E; (**f**) Specimen F; (**g**) Specimen G.

Table 8. Finite-element simulation results.

Specimen ID	σ_{max} (MPa)	*FE/Test*	$\sigma_{max,t}$ (MPa)	*FE/Test*	Hysteresis Envelope Area (MPa)	*FE/Test*	$N_{B,S}$	$N_{B,T}$
A	611.6	0.981	535.93	1.071	218.98	1.114	10	10
B	545.4	0.933	517.52	1.146	143.14	1.130	6	7
C	442.7	0.851	471.84	1.070	103.02	1.145	5	5
D	438.3	0.870	501.25	1.123	90.40	1.205	3	3
E	426.3	0.954	478.82	1.122	105.94	1.274	3	3
F	551.7	0.947	494.38	1.161	83.65	1.146	5	7

$N_{B,S}$—the number of cycles when the finite-element buckling phenomenon occurs; $N_{B,T}$—the number of cycles when the test buckling phenomenon occurs.

4.4. Parametric Study and Analysis

In order to further study the influencing factors on the hysteretic performance of FCCC-R specimens to compensate for the shortcomings of incomplete and superficial testing in this study and based on the previous specimen C finite-element model, we carried out an expanded parametric study. The number of winding layers *L* and winding angles *J* of the GFRP strips, as well as the rebar-position eccentricity *e/R*, were used as analysis parameters for this study. The number of winding layers refers to the total number of layers of GFRP strips wrapped on the outside of the high-strength mortar. The winding angle is the included angle between the fiber direction of the GFRP strips and the surrounding direction of the FCCC-R specimens. The rebar-position eccentricity refers to the ratio of the distance from the center of the rebar section to the center of the GFRP tube to the constraint radius. The value ranges of the parameters are listed in Table 9. For purposes of comparison and analysis, the skeleton curves of the FCCC-R models with different influence parameters are displayed to roughly reflect their hysteretic performance. The simulation results are shown in Figure 14.

Table 9. Values of the parameters considered in the expanded parametric study.

Parameter	Values
Winding layers, L	1, 3, 5
Winding angle, J	30°, 45°, 60°
Rebar-position eccentricity, e/R	0.15, 0.22, 0.3

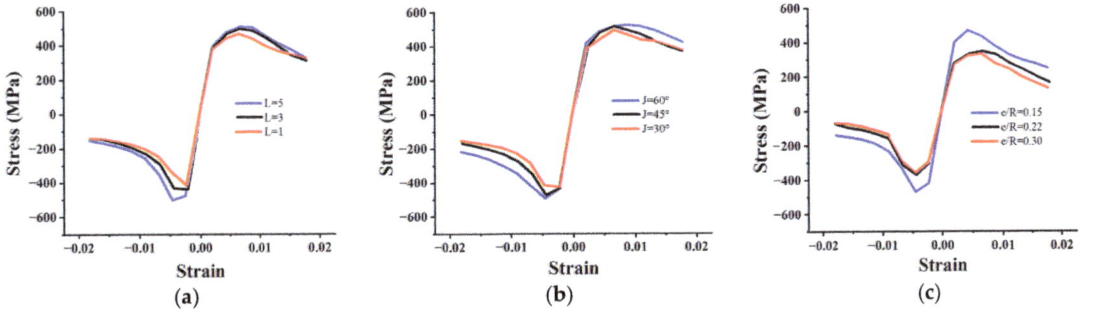

Figure 14. Skeleton curves of FCCC-R with different influence parameters. (**a**) Winding layers; (**b**) Winding angle; (**c**) Rebar-position eccentricity.

Based on the simulation results, it is concluded that the hysteretic performance of FCCC-R increases to a certain extent with the increase in the number of winding layers and angles of GFRP strips. This is due to the fact that when 0° winding is used, the GFRP strips of the FCCC-R specimens are only affected by the adhesive force of the fibers. However, the tensile bearing capacity of the fibers is greater than the bonding force between them; therefore, as the winding angle increases, the tensile properties of the fibers begin to play an increasingly important role, continuously improving the hysteretic performance of FCCC-R. Similarly, as the number of winding layers on the GFRP strips increases, the superposition of the effectiveness of multilayer GFRP strips also enhances the constraint effect of restraint components. This leads to an improvement in the constraint effect of external components on rebar. As the constraint effect of external components increases, rebar can further exert its mechanical properties, resulting in an increase in peak stress in compression and tension at each cycle of loading. There are some similarities between this enhancement mechanism and increasing the diameter of the external constraint to improve the hysteretic performance of rebar. However, due to the much larger tensile bearing capacity in the fiber direction compared to mortar, on the tensile side of the rebar, GFRP strips have a greater effect on the restraint of the rebar.

Contrary to the above conclusions, the hysteretic behavior of FCCC-R specimens decreases continuously with the increase in rebar-position eccentricity. The reason for this is that the existence of eccentricity leads to the existence of a weak part on one side of the restraint component, which weakens the confined effectiveness of the eccentric side in terms of restraining the rebar capacity when subjected to cyclic loading, resulting in a reduction in the hysteretic performance of the FCCC-R specimens. When the eccentricity reaches 0.22, the hysteretic properties of the rebar will be clearly reduced.

5. Conclusions

This study examined the hysteretic behavior of FRP-confined concrete core-encased rebar (FCCC-R) under cyclic loading. The main parameters for evaluating the hysteretic properties of FCCC-R specimens are the loading lengths and constraint diameters. Additionally, the study investigated the effects of different loading systems on these properties. The elongation and mechanical properties of specimens under different cyclic loading systems were compared with those of ordinary rebar under monotonic tension. Finite-element simulation and corresponding expansion parameter analysis were also carried out. The following conclusions are drawn from the study results:

(i) Compared to ordinary rebar, FCCC-R specimens exhibit superior hysteretic performance, as evidenced by their higher compression peak load and fracture strain, increased number of load cycles, larger hysteresis envelope area, and lower fracture stress. Additionally, the buckling phenomenon of FCCC-R specimens during loading is delayed compared to that of ordinary rebar due to the confinement provided by the high-strength mortar and GFRP strips;

(ii) Under the same slenderness ratio, the hysteretic performance of FCCC-R was improved to varying degrees with the increase in the constraint diameter. However, the hysteretic behavior of the specimens with a small constraint diameter was not significantly different from that of the ordinary rebar. Therefore, with the increase in slenderness ratio, it was necessary to appropriately increase the constraint diameter in order to achieve a significant improvement;

(iii) The hysteretic properties of the specimens under tension–compression cyclic loading were consistent with those under tension–tension cyclic loading. The hysteretic properties of the specimens decreased with the increase in loading length. However, due to the small cumulative damage of rebar under tension–tension cyclic loading, its hysteretic performance was better than that under tension–compression cyclic loading;

(iv) Through comparison of the elongation of specimens under different loading systems, it was observed that the FCCC-R specimens exhibited greater elongation under tension–tension cyclic loading than those under tension–compression cyclic loading. In addition, under any cyclic loading system, the fracture elongation of the FCCC-R specimens was greater than that of ordinary rebar but still lower than the elongation of the maximum bearing capacity of ordinary rebar under monotonic tension. Under cyclic loading, specimen buckling occurred earlier but FCCC-R effectively delayed the phenomenon of rebar buckling. Moreover, under tension–compression cyclic loading, the maximum compressive bearing capacity of the FCCC-R specimens was improved compared to monotonic compressive loading;

(v) Finite-element models of different FCCC-Rs were established by using ABAQUS software for numerical analysis. The model was validated through comparison with experimental results. Except for a small amount of data with a discrepancy of 20%, the data discrepancy was approximately 10%. Therefore, the numerical simulation results agreed well with the experimental results. Based on these results, expanded parametric studies were also carried out to analyze the effects of different winding layers, winding angles of GFRP strips, and rebar-position eccentricity on the hysteretic properties of FCCC-R. It was concluded that the hysteretic performance of FCCC-R increased with the increase in the number of GFRP winding layers and winding angles. The increase in rebar-position eccentricity continuously reduced the hysteretic performance of FCCC-R;

(vi) According to the research in this article, when designing FCCC-R, it is advisable to adopt a larger wrapping angle of GFRP strips when the concrete expansion is small. The constraint diameter of the constraint component cannot be too small. In this article, it is recommended that the minimum area ratio of rebar to constrained components be 0.54. In addition, reducing the rebar-position eccentricity as much as possible is conducive to improving the ground hysteretic performance of FCCC-R.

Author Contributions: Conceptualization, J.L.; methodology, J.L.; validation, T.M.; formal analysis, T.M.; investigation, H.H.; resources, J.L.; data curation, C.W.; writing—original draft preparation, T.M.; writing—review and editing, J.L.; visualization, T.M.; supervision, W.H.; project administration, J.L.; funding acquisition, J.L. All authors have read and agreed to the published version of the manuscript.

Funding: This research was funded by the National Natural Science Foundation of China (Grant No. 51978596).

Institutional Review Board Statement: Not applicable.

Informed Consent Statement: Not applicable.

Data Availability Statement: Data sharing is not applicable to this article.

Conflicts of Interest: The authors declare no conflict of interest.

References

1. Chang, X.; Yang, L.; Zong, L.; Zhao, M.H.; Yin, F. Study on cyclic constitutive model and ultra low cycle fracture prediction model of duplex stainless steel. *J. Constr. Steel Res.* **2019**, *152*, 105–116. [CrossRef]
2. Nip, K.H.; Gardner, L.; Davies, C.M.; Elghazouli, A.Y. Extremely low cycle fatigue tests on structural carbon steel and stainless steel. *J. Constr. Steel Res.* **2010**, *66*, 96–110. [CrossRef]
3. Lukkunaprasit, P.; Tangbunchoo, T.; Rodsin, K. Enhancement of seismic performance of reinforced concrete columns with buckling-restrained reinforcement. *Eng. Struct.* **2011**, *33*, 3311–3316. [CrossRef]
4. Qiu, J.L.; Gong, J.X. Research on buckling of longitudinal bars in RC columns under earthquake excitation: State of the art. *China Civ. Eng. J.* **2016**, *49*, 50–62. (In Chinese)
5. Shi, Q.X.; Ma, L.C.; Wang, Q.W.; Wang, B.; Yang, K. Seismic performance of square concrete columns reinforced with grade 600 MPa longitudinal and transverse reinforcement steel under high axial load. *Structures* **2021**, *32*, 1955–1970. [CrossRef]
6. Su, J.S.; Wang, J.J.; Bai, Z.Z.; Wang, W.B.; Zhao, D.X. Influence of reinforcement buckling on the seismic performance of reinforced concrete columns. *Eng. Struct.* **2015**, *103*, 174–188. [CrossRef]
7. Massone, L.M.; López, E.E. Modeling of reinforcement global buckling in RC elements. *Eng. Struct.* **2014**, *59*, 484–494. [CrossRef]
8. Zheng, J.F. Experiment Studies on Cyclic Behavior of Reinforcing Bars including Buckling. Master's Thesis, Chongqing University, Chongqing, China, 2012.
9. Dhakal, R.P.; Su, J. Design of transverse reinforcement to avoid premature buckling of main bars. *Earthq. Eng. Struct. Dyn.* **2018**, *47*, 147–168. [CrossRef]
10. Hashemi, S.S.; Vaghefi, M.; Hemmat, M. Evaluation the effects of stirrup spacing and buckling of steel reinforcing bars on the capacity of RC columns. *Scientia Iranica* **2017**, *25*, 1140–1151. [CrossRef]
11. Pereiro-Barceló, J.; Bonet Senach, J.L.; Albiol-Ibáñez, J.R. Required tie spacing to prevent inelastic local buckling of longitudinal reinforcements in RC and FRC elements. *Eng. Struct.* **2018**, *160*, 328–341. [CrossRef]
12. Mitra, D.C.; Bindhu, K.R. Seismic performance of RC short columns with buckling restraint reinforcement. *Mag. Concr. Res.* **2019**, *71*, 163–174. [CrossRef]
13. Ruangrassamee, A.; Sawaroj, A. Seismic enhancement of reinforced-concrete columns by rebar-restraining collars. *J. Earthq. Tsunami* **2013**, *6*, 1250015.1–1250015.19. [CrossRef]
14. Ye, L.P.; Feng, P. Applications and development of fiber-reinforced polymer in engineering structures. *China Civ. Eng. J.* **2006**, *3*, 24–36. (In Chinese)
15. Rashid, S.M.P.; Bahrami, A. Structural Performance of Infilled Steel–Concrete Composite Thin-Walled Columns Combined with FRP and CFRP: A Comprehensive Review. *Materials* **2023**, *16*, 1564. [CrossRef]
16. Baena, M.; Jahani, Y.; Torres, L.; Barris, C.; Perera, R. Flexural Performance and End Debonding Prediction of NSM Carbon FRP-Strengthened Reinforced Concrete Beams under Different Service Temperatures. *Polymers* **2023**, *15*, 851. [CrossRef]
17. Zhu, H.; Li, C.; Cheng, S.; Yuan, J. Flexural Performance of Concrete Beams Reinforced with Continuous FRP Bars and Discrete Steel Fibers under Cyclic Loads. *Polymers* **2022**, *14*, 1399. [CrossRef] [PubMed]
18. Miao, K.; Wei, Y.; Zhang, X.; Zheng, K.; Dong, F. Performance of Circular Concrete-Filled FRP-Grooved Steel Composite Tube Columns under Axial Compression. *Polymers* **2021**, *13*, 3638. [CrossRef] [PubMed]
19. Tahir, M.; Wang, Z.Y.; Ali, K.M.; Isleem, H.F. Shear behavior of concrete beams reinforced with CFRP sheet strip stirrups using wet-layup technique. *Structures* **2019**, *22*, 43–52. [CrossRef]
20. Feng, P.; Zhang, Y.H.; Bai, Y.; Ye, L.P. Strengthening of steel members in compression by mortar-filled FRP tubes. *Thin-Walled Struct.* **2013**, *64*, 1–12. [CrossRef]
21. Wang, Z.Y.; Feng, P.; Zhao, Y.; Yu, T. FRP-confined concrete core-encased rebar for RC columns: Concept and axial compressive behavior. *Compos. Struct.* **2019**, *222*, 110915.1–110915.13. [CrossRef]
22. Lu, J.Z.; Huang, H.; Li, Y.K.; Mou, T. Experimental and Numerical Investigation of Axial Compression Behaviour of FRP-Confined Concrete-Core-Encased Rebar. *Polymers* **2023**, *15*, 828. [CrossRef]

23. Wang, Z.Y.; Li, Z.Y.; Feng, P. Cyclic axial compressive performance of the RC columns reinforced with FRP confined concrete core encased rebar. *Eng. Struct.* **2023**, *274*, 115166. [CrossRef]
24. Hu, L.; Feng, P.; Yang, J.Q.; Li, Z.Y. Shear behaviour of FRP-confined concrete/UHPC core-encased rebar: Experimental investigation and design method. *Eng. Struct.* **2023**, *284*, 115951. [CrossRef]
25. Vizentin, G.; Vukelic, G. Prediction of the Deterioration of FRP Composite Properties Induced by Marine Environments. *J. Mar. Sci. Eng.* **2022**, *10*, 510. [CrossRef]
26. Shen, C.; Feng, P.; Sun, L.; Li, Z.Y. Experimental study of RC columns reinforced with FCCC-R under eccentric load. In Proceedings of the 29th National Conference on Structural Engineering, Wuhan, China, 16 October 2020.
27. Dong, Z.Q.; Wu, G. Research progress on durability of FRP bars reinforced concrete structures. *China Civ. Eng. J.* **2019**, *52*, 1–19. (In Chinese)
28. Gao, X.D.; Shao, Y.B.; Chen, C.; Feng, R.; Zhu, H.M.; Li, T. Hysteresis behavior of EQ56 high strength steel: Experimental tests and FE simulation. *J. Constr. Steel Res.* **2023**, *201*, 107730. [CrossRef]
29. Dusicka, P.; Itani, A.M.; Buckle, I.G. Cyclic response of plate steels under large inelastic strains. *J. Constr. Steel Res.* **2007**, *63*, 156–164. [CrossRef]
30. Zhou, F.; Chen, Y.Y.; Wu, Q. Dependence of the cyclic response of structural steel on loading history under large inelastic strains. *J. Constr. Steel Res.* **2015**, *104*, 64–73. [CrossRef]
31. Chen, F.J.; Yi, W.J. Study of high-strength steel under variable amplitude cyclic-load test. *Ind. Constr.* **2016**, *46*, 154–158. (In Chinese)
32. Chaht, F.L.; Mokhtari, M.; Benzaama, H. Using a Hashin criteria to predict the damage of composite notched plate under traction and torsion behavior. *Fract. Struct. Integr.* **2019**, *50*, 331–341.
33. Son, J.; Fam, A. Finite element modeling of hollow and concrete-filled fiber composite tubes in flexure: Model development. Verif. Investig. Tube Parameters. *Eng. Struct.* **2008**, *30*, 2656–2666. [CrossRef]
34. Chaboche, J.L. Time-independent constitutive theories for cyclic plasticity. *Int. J. Plast.* **1986**, *2*, 149–188. [CrossRef]
35. Dai, G.X.; Wang, F.; Shi, G.; Wang, Y.Q.; Shi, Y.J. Comparison of monotonic and cyclic performances of structural steel q345 and q460. *Ind. Constr.* **2016**, *42*, 13–17. (In Chinese)
36. Wang, Y.B.; Li, G.Q.; Sun, X.; Chen, S.W.; Hai, L.T. Evaluation and prediction of cyclic response of Q690D steel. *Proc. Inst. Civ. Eng.-Struct. Build.* **2017**, *170*, 788–803. [CrossRef]
37. Hai, L.T.; Sun, F.F.; Zhao, C.; Li, G.Q.; Wang, Y.B. Experimental cyclic behavior and constitutive modeling of high strength structural steels. *Constr. Build. Mater.* **2018**, *189*, 1264–1285. [CrossRef]

polymers

MDPI

Article

Experimental and Numerical Investigation of Axial Compression Behaviour of FRP-Confined Concrete-Core-Encased Rebar

Jingzhou Lu *, Han Huang *, Yunkai Li and Tong Mou

School of Civil Engineering, Yantai University, Yantai 264005, China
* Correspondence: jingzhoulu@ytu.edu.cn (J.L.); hhzgq2580@s.ytu.edu.cn (H.H.)

Abstract: The axial compression behaviour of fibre-reinforced polymer (FRP)-confined concrete-core-encased rebar (FCCC-R) was investigated by performing monotonic axial compression tests on seven groups of FCCC-R specimens and three groups of pure rebar specimens. The research parameters considered were the FRP winding angle (0°, ±45°, and 90°), number of layers (2, 4, and 6 layers), and slenderness ratio of specimens (15.45, 20, and 22.73). The test results showed that FCCC-R's axial compression behaviour improved significantly compared with pure rebar. The axial load–displacement curves of the FCCC-R specimens had a second ascending branch, and their carrying capacity and ductility were enhanced substantially. The best buckling behaviour was observed for the FRP winding angle of 90°. The capacity and ductility of the specimens were positively related to the number of FRP-wrapped layers and inversely related to the slenderness ratio of the specimens. A finite element model of FCCC-R was constructed and agreed well with the test results. The finite element model was used for parametric analysis to reveal the effect of the area ratio, FRP confinement length, internal bar eccentricity, and mortar strength on the axial compression behaviour of FCCC-R. The numerical results showed that the area ratio had the most significant impact on the axial compression behaviour of FCCC-R. The confinement length of the FRP pipe and internal bar eccentricity had similar effects on the axial compression behaviour of FCCC-R. Both of them had a significant impact on the second ascending branch, with the post-peak behaviour exhibiting minimal differences. The influence of mortar strength on the axial compression behaviour of FCCC-R was observed to be minimal.

Keywords: FRP-confined concrete-core-encased rebar; buckling; monotonic loading; axial compressive behaviour; finite element analysis

1. Introduction

Located between the circum-Pacific seismic belt and the Eurasian seismic belt, China is squeezed by the Pacific plate, Indian Ocean plate, and Philippine Sea plate, resulting in frequent earthquake activities. Laboratory evidence and earthquake damage observations indicate that the buckling of longitudinal steel bars is a vital damage state of reinforced concrete (RC) columns when subjected to ever-increasing side-sway inelastic deformations [1–3].

For RC columns with insufficient lateral reinforcement, a large transversal reinforcement distance and large axial compression ratio and axial compressive force combined with a bending moment will produce a large strain at the section edge, and large tensile strains followed by high compression will cause the concrete to crush or spall. When the cover concrete spalls, the longitudinal reinforcement loses the support of the surrounding concrete, resulting in inelastic buckling due to insufficient restraint. The reinforcement will undergo large cyclic deformation after buckling. Early fracture of the reinforcement is likely to occur under low-cycle fatigue, resulting in the loss of the lateral bearing capacity and the collapse of the RC column [4]. Therefore, there is a need to conduct research to develop ways to increase the rebar ductility and prevent the longitudinal rebar from premature buckling. Developing such techniques can assist in the construction of structures with elevated seismic resistance performance.

To prevent premature buckling of the longitudinal rebar in reinforced concrete columns, researchers have proposed many approaches, among which the appropriate arrangement of lateral reinforcement in reinforced concrete columns has been widely studied and was accepted by concrete construction design criteria [5–7]. However, extra lateral reinforcement might make the steel rebar overcrowded, which is not only inconvenient for construction in the beam–column joint area but also makes it difficult to cast and vibrate concrete evenly to achieve the expected strength [4]. In addition, a hybrid reinforcing component, called a steel restraint ring, which uses a steel pipe to restrain the rebar, has also been developed to prevent the buckling of the longitudinal rebar [8–10]. The component works better on columns with slimmer rebar compared with regular reinforced concrete columns. With the development of fibre-reinforced polymer (FRP), it was realised that FRP had excellent properties such as high strength, fatigue resistance, and corrosion resistance. In recent years, a lot of research results have been achieved on FRP instead of steel-tube-confined concrete, and the confined specimens all showed significant compressive performance enhancement [11–21]. Feng et al. [22] proposed a concept in which a slender steel component was introduced into a mortar-filled FRP tube. Axial compression tests of a series of specimens showed that the specimen had better axial compressive behaviour than a pure steel component. Wang et al. [4] proposed a similar concept: FRP-confined concrete-core-encased rebar (FCCC-R). Their research showed that the constraining effect of concrete and the FRP tube effectively prevented premature buckling of the internal rebar. On the one hand, for a single FCCC-R specimen, the restraint effect of FRP improves the compressive performance of concrete or mortar and prevents the concrete cover from spalling. Under the support of the outer concrete, the buckling performance of the internal reinforcement has been greatly improved. In addition, under the double reinforcement of the FRP pipe and mortar, the section inertia moment of the internal reinforcement increases, which enhances the axial compression performance of the specimen. On the other hand, for reinforced concrete RC columns equipped with FCCC-R, the high-strength concrete or mortar filled in the FRP pipe improves the axial compressive strength of the column. In addition, the confinement of the FRP pipe can effectively improve the strength of confined concrete [23]. Under the action of seismic load, after the concrete surrounding FCCC-R in the RC column is crushed under compression, FCCC-R can still be used as a high-strength skeleton to bear the force alone, which significantly improves the anti-seismic performance of the column. In their study, a few horizontal cracks appeared on the mid-span side of the specimens. This was because the FRP winding direction adopted for all FCCC-R specimens was close to 90° to the cross-sectional direction of the specimen, which only provided better hoop constraints. Once the FRP pipe was bent, cracks easily developed on the side, resulting in the failure of the entire specimen. Therefore, it is necessary to study the effect of the FRP winding angle on the axial compression behaviour of FCCC-R.

Using the concept of FCCC-R proposed by Wang et al. [4], this study further investigated and analysed its buckling stability. Monotonic axial compression tests of seven groups of FCCC-R specimens and three control groups of pure rebar specimens were performed to analyse the axial compression behaviour and failure mode of specimens with different winding angles, winding thicknesses, and slenderness ratios. Furthermore, a finite element model was constructed to verify the test results. The finite element model was also used to perform parametric analyses to reveal the effects of the area ratio, FRP pipe confinement length, internal bar eccentricity, and mortar strength on the axial compression behaviour of FCCC-R.

2. Overview of Test

2.1. Specimen Design

The configuration of the FCCC-R specimen is shown in Figure 1. The exterior of the specimen was an FRP pipe, and the internal reinforcement was placed at the centre of the cross section of the FRP pipe. High-strength concrete or mortar was filled in the middle of the FRP pipe and internal rebar.

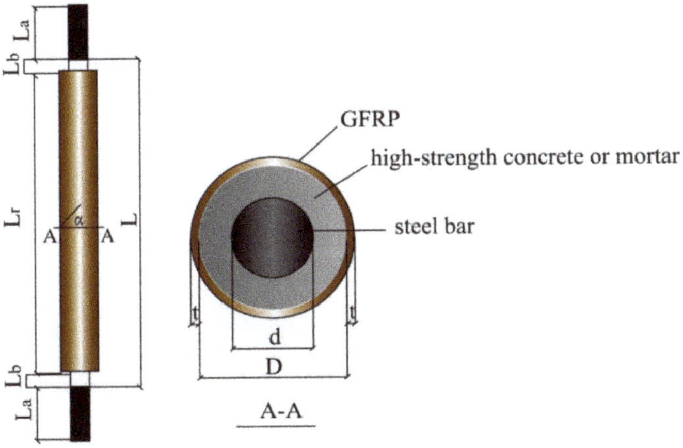

Figure 1. Configuration of fibre-reinforced-polymer-confined concrete-core-encased rebar (FCCC-R) specimens.

The length of the FRP pipe was defined as the confinement length L_r. For all FCCC-R specimens, a reserve length of 90 mm at each end of the steel bar was not constrained by the FRP pipe, and out of this reserve length, L_a = 70 mm was the supporting length in the test to ensure that both ends were fixed; L_b = 20 mm was set as the length between the end of the test fixture and the end of the FRP tube to ensure that the axial load was applied only to the internal rebar. $L = L_r + 2L_b$ was the loading length of the specimen. The fibre winding angle α was defined as the angle between the fibre and the cross section of the specimen, and it was positive in the counter-clockwise direction and negative in the clockwise direction. Furthermore, the steel rebar's diameter d was 22 mm, the core concrete's diameter D was 40 mm, and the thickness of the FRP pipe's wall, t, was 2 mm.

2.2. Specimen Preparation

The preparation of the FCCC-R specimen was divided into three steps: mould making; mortar pouring, mould removal, and maintenance; and fibreglass cloth wrapping. These steps are described below.

Mould making: Made of a ready-made PVC pipe, the specimen mould was divided into the middle and the base parts, as shown in Figure 2a. To study the buckling performance of FCCC-R at different slenderness ratios, we intercepted 300, 400, and 500 mm long PVC pipes. A lubricant was then applied to these pipes to place them in the middle of the mould. A 90 mm long PVC pipe was cut to form the mould base because FCCC-R required 90 mm long rebar exposed at both ends, with the top being left outside. One end of the rebar was wrapped with soft plastic and placed in the base to ensure that the rebar was in the centre position and to prevent the mortar from being exposed at the bottom of the mould. Subsequently, the middle of the mould was connected to the base with adhesive tape, ready for mortar pouring.

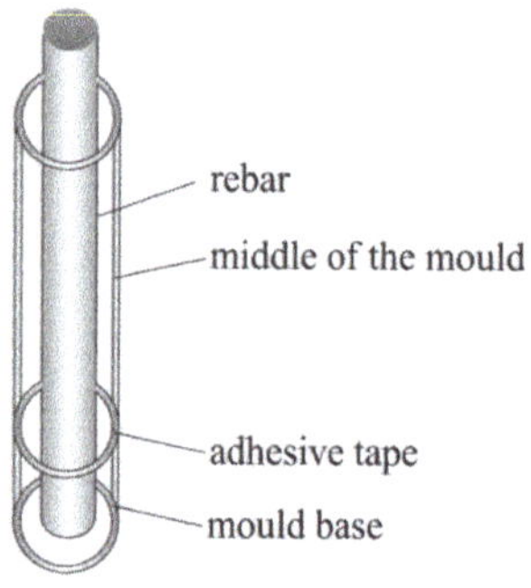

(a) Schematic view of the mould.

(b) The reinforced FCCC-R.

Figure 2. Schematic view of the mould and reinforced FCCC-R specimens.

Mortar pouring, mould removal, and maintenance: Mortar and water were taken in a mortar mixer and continuously mixed. The mixture was poured into the mould when the mortar fluidity met the requirement. The sidewall of the mould was tapped while pouring the mixture, and a hollow circular iron plate was used to position the upper part of the rebar. Three cubic mortar specimens were poured simultaneously. After completing this, the mould was removed, and the mortar specimens were placed in a concrete curing room for 28 days.

Fibreglass cloth wrapping: Before the mortar specimens were wrapped in fibreglass cloth, sandpaper was used to remove floating ash on their surface, and the exposed steel bars at both ends were wrapped with plastic film. Then, fibre cloth with a lap length appropriate for the test study parameters was cut. A binder was evenly applied to the outer surface of the mortar and rested for a period of time. After the outer surface of the mortar had dried, the fibre cloth was pasted on it. Finally, the specimens were placed in a dry environment for seven days for curing and reinforcing. The reinforced FCCC-R specimens are shown in Figure 2b.

2.3. *Test Materials*

2.3.1. Rebar

In this test, HRB400 rebar with a diameter of 22 mm was used. Tensile tests were performed on it according to GB/T228.1-2010, Tensile Test of Metal Materials at Room Temperature Part 1: Greenhouse Test Method. The measured material property indexes of the rebar are presented in Table 1.

Table 1. Material property indexes of rebar.

Rebar Type	Yield Strength (MPa)	Yield Strain	Tensile Strength (MPa)	Elastic Modulus (GPa)	Poisson's Ratio
HRB400	409	0.00207	566	196.7	0.3

2.3.2. Mortar

In this test, in order to better cast and form FCCC-R, mortar with the same strength as concrete but with better fluidity was selected. The mortar material was high-strength non-shrink grouting material produced by Qingdao ZhuoNengda Construction Technology Co., Ltd. (Qingdao, China) Samples required for the material performance test were prepared according to GB/T50081-2019, Standard for test methods for physical and mechanical properties of concrete, and the method presented in the literature [4]. After preparation, the specimens were placed in a curing room at a constant temperature and humidity. After 24 h, the mould was shifted to a standard concrete curing room, which was kept for 28 days. The WE-1000 hydraulic universal testing machine was used to perform the loading test on the mortar specimens, and the cubic compressive strength of the mortar specimens was measured to be 43.59 MPa. In order to better perform the finite element analysis of the specimen, the data were converted into cylinder axial compressive strength. The material property indexes of the mortar are presented in Table 2.

Table 2. Material property indexes of mortar.

Coupon Shape	Axial Compressive Strength (MPa)	Elastic Modulus (GPa)	Poisson's Ratio
Cylinder	34.44	27.76	0.21

2.3.3. FRP

FRP was made of unidirectional glass fibre cloth and epoxy resin glue produced by Hai Ning Anjie Co., Ltd. After the fibre sheet was prepared, the tensile test was performed according to GB/T3354-2014, Test method for tensile properties of oriented fibre-reinforced polymer matrix composites. Furthermore, the longitudinal and horizontal shear test was performed according to GB/T3856-2005, Longitudinal and transverse shear test method for polymer matrix composites. The mechanical property indexes of composite materials are presented in Table 3.

Table 3. Mechanical property indexes of composite materials.

Longitudinal Tensile Strength (MPa)	Transverse Tensile Strength (MPa)	Shear Strength (MPa)	Longitudinal Elastic Modulus (GPa)	Transverse Elastic Modulus (GPa)	Shear Modulus (GPa)	Poisson's Ratio
760.2	55.1	217.1	45.1	2.7	14.9	0.23

2.4. *Test Method*

In order to study the effects of the FRP winding angle, winding thickness, and slenderness ratio of the specimen on the buckling performance of FCCC-R under an axial compressive load, the following steps were followed. The number of FRP constraint layers was divided into internal and external layers, with the number of both layers being equal. All FRP external layers were circumferentially wrapped, that is, $\alpha = 0°$, to protect the specimen from developing longitudinal cracks because of the insufficient hoop constraint. In this part, the internal layer winding angles were varied to study the effect of the FRP winding angle, the number of FRP layers was varied to study the effect of the winding thickness, and the length of the FRP pipe was varied to study the effect of the slenderness ratio of the specimen.

Monotonic axial compression tests were performed on seven groups of FCCC-R specimens and three groups of pure rebar specimens, with each group of specimens containing three identical specimens to avoid behavioural deviation caused by the rebars' initial defect and initial load eccentricity. The monotonic loading specimens are listed in Table 4. The nomenclature of the specimen groups is as follows: for FCCC-R specimens, the group name begins with "F", followed by a number indicating the length of the FRP restraint. The number after the slash indicates the loading length of the specimen. The numbers '2', '4', and '6' indicate the number of winding layers of FRP, and the letters 'a', 'b', and 'c' indicate fibre winding angles of '0°', '±45°', and '90°' for the internal layer, respectively. For normal rebar, the group name starts with "R" and the number after it indicates the loading length of the rebar.

Table 4. Monotonic loading specimens.

Specimen ID	Confinement Length (mm)	Loading Length (mm)	Layers	Confinement Thickness (mm)	Angle (from Inside to Outside)
F-400/440-4-a	400	440	4	2.8	0°0°0°0°
F-400/440-4-b	400	440	4	2.8	−45° + 45°0°0°
F-400/440-4-c	400	440	4	2.8	90°90°0°0°
F-400/440-2-c	400	440	2	1.4	90°0°
F-400/440-6-c	400	440	6	2.4	90°90°90°0°0°0°
F-300/340-4-c	300	340	4	2.8	90°90°0°0°
F-500/540-4-c	500	540	4	2.8	90°90°0°0°
R-340	–	340	–	–	–
R-440	–	440	–	–	–
R-540	–	540	–	–	–

All the specimens were pressurised on an SDS500 electro-hydraulic servo dynamic and static universal testing machine, and the maximum range of the testing machine was 500 kN. In the test, after switching to the displacement loading mode, we set the loading amplitude to 1 mm/min and recorded the load–displacement relationship of the specimens.

The test setup of the FCCC-R specimen is illustrated in Figure 3. During the axial compression test, the test setup was calibrated vertically to ensure that the specimen could be damaged under axial compression and reduce the eccentric pressure damage as much as possible. The two fixtures of the testing machine provided fixed-end conditions during the test. The mean value of the result obtained by the two vertical LVDTs was used to eliminate the influence of the slight rotation of the clamps that occurred during the test.

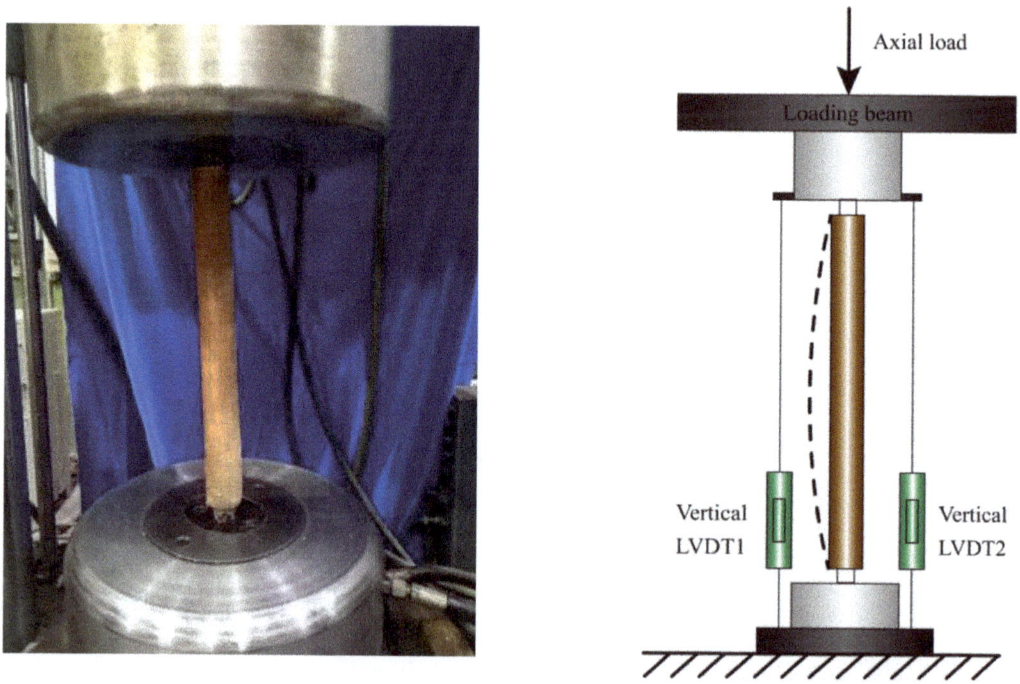

(**a**) Scene diagram. (**b**) Schematic view.

Figure 3. Test setup of FCCC-R specimen.

3. Test Result and Analysis

3.1. Test Phenomena and Mechanism Analysis

3.1.1. Rebar Buckling

In the initial stage, the load increased monotonically with the loading time. Soon, it reached the peak, and it subsequently decreased rapidly. At this time, it was observed that the steel bar was rapidly bent, resulting in the "buckling" phenomenon, and the test was terminated to ensure safety. The specimens R-340, R440, and R540 all buckled before reaching the yield load. The reason is that rebar buckles before reaching the yield load when the slenderness ratio exceeds 9 or 10, and the slenderness ratios of all three groups of specimens exceeded 10.

3.1.2. Mechanism Analysis of FCCC-R Buckling Failure

In the beginning, the FCCC-R specimens remained elastic, and the relationship between the axial load and displacement was basically linear. As the force acting on the FRP cloth and the axial strain value was relatively small, the restraint effect was not obvious. As the axial load increased, the strain of the FRP cloth started increasing significantly, with the strain rate gradually accelerating. The acceleration was because the appearance and development of concrete micro-cracks at this time increased the lateral deflection of the specimen, resulting in an increase in the horizontal restraint of the FRP cloth. When the stiffness of the specimen decreased to a smaller value, the load–displacement curve exhibited a second rising stage where it reached a peak. At this stage, the FRP-confined area became slightly curved. Shortly after the peak load, the FCCC-R specimen buckled, the lateral displacement increased significantly, and transverse cracks appeared rapidly with an obvious cracking sound. The FRP outer layer fractured suddenly, and the bearing capacity of the specimen plummeted [24]. Simultaneously, the axial load decreased rapidly, entering the softening stage. Finally, the specimen was destroyed due to the loss of carrying capacity and excessive lateral displacement.

The overall buckling failure mode of the FCCC-R specimens is shown in Figure 4. The global buckling failure occurred in all specimens, and circumferential cracks appeared in the middle of the specimens. During testing, it was found that the fibre winding angle affected the number and morphology of cracks. Multiple circumferential cracks appeared in the middle of the specimens with an inner layer winding angle of 0°, while the middle of the specimens with inner layer fibre winding angles of ±45°and 90° mainly developed a circumferential crack, with a few smaller circumferential cracks appearing on both sides of the main crack. The explanation for this is as follows. When the outer and inner fibres of the FCCC-R specimens were all circumferentially wound, the entire FRP pipe had a strong circumferential restraint, and the longitudinal tension came only from the adhesive force associated with the fibre glue. Once the specimen buckled, multiple circumferential cracks were rapidly formed. When the inner fibre winding angle was ±45°or 90°, the entire FRP pipe not only produced a certain hoop restraint but also provided a certain longitudinal tension, which reduced the formation of hoop cracks.

On some specimens, diagonal cracks appeared at the edge of the mortar. Local failure diagrams of the FCCC-R specimens are shown in Figure 5. During the loading process of the specimens, the fibres at the ends of one F-400/440-2-c specimen and one F-300/340-4-c specimen were torn longitudinally, and the load dropped rapidly before reaching the ideal value. The reason was that the rebar was incorrectly positioned. It was not placed in the middle of the specimen during specimen production, resulting in insufficient local hoop restraint which caused the formation of longitudinal cracks.

(a) F-400/440-4-a. (b) F-400/440-4-b. (c) F-400/440-4-c. (d) F-400/440-2-c.

(e) F-400/440-6-c. (f) F-300/340-4-c. (g) F-500/540-4-c.

Figure 4. Global buckling failure mode of FCCC-R specimens.

(a) Multiple circumferential cracks. (b) A circumferential crack. (c) Diagonal crack.

Figure 5. Local failure diagrams for FCCC-R specimens.

3.2. Test Result Comparison and Analysis

3.2.1. Summary of Test Results

The buckling test results of the specimens are presented in Table 5. P_{max} is the buckling peak load of the specimen, $P_{max,s}$ is the buckling peak load of the rebar with the same slenderness ratio, U_{max} is the axial displacement corresponding to P_{max}, and F_y is the yield load of the steel rebar. The values of $P_{max}/P_{max,s}$ are greater than 1, and it can be seen that the FCCC-R specimens improve the compressive performance of steel bars to different extents. The values of P_{max}/F_y showed that the pure steel rebars suffered buckling failure before reaching the yield strength, while the steel bars of the FCCC-R specimens yielded when buckling instability occurred.

Table 5. Test results for specimens.

Specimen ID	L/d	P_{max} (kN)	U_{max} (mm)	P_{max}/F_y	$P_{max}/P_{max,s}$
F-400/440-4-a	20	180.16	2.67	1.16	1.31
F-400/440-4-b	20	190.90	3.55	1.23	1.39
F-400/440-4-c	20	203.69	4.12	1.31	1.48
F-400/440-2-c	20	178.83	2.64	1.15	1.30
F-400/440-6-c	20	231.00	5.68	1.49	1.68
F-300/340-4-c	15.45	238.17	7.05	1.53	1.62
F-500/540-4-c	22.73	178.55	2.56	1.15	1.39
R-340	15.45	146.63	0.82	0.94	
R-440	20	137.63	0.84	0.89	
R-540	22.73	128.42	0.89	0.83	

3.2.2. Analysis of Effect of Winding Angle on FCCC-R Axial Compression Performance

It is evident from Table 5 and Figure 6a that when the thickness of the fibre layer and the slenderness ratio were constant, all FCCC-R specimens, irrespective of the fibre winding angle, had a second ascending branch, and their carrying capacity and ductility appeared to be better than those of a pure steel bar with the same slenderness ratio. The peak loads of the FCCC-R specimens with inner fibre winding angles of 0°, ±45°, and 90° were 31%, 39%, and 48% higher than those of the corresponding pure steel bars, respectively. The axial displacements corresponding to the peak loads were 218%, 323%, and 390% higher than those of the corresponding pure steel bars, respectively.

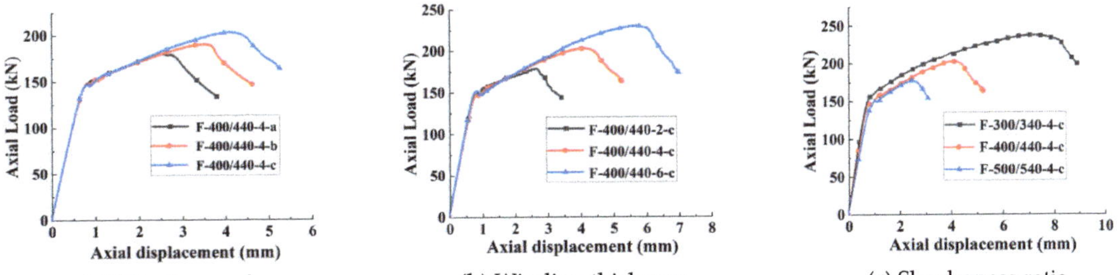

(**a**) Winding angle. (**b**) Winding thickness. (**c**) Slenderness ratio.

Figure 6. Comparison of load–displacement relationships between different groups.

Therefore, when the thickness of the fibre layer and the slenderness ratio were constant, the optimal winding angle of the inner layer fibre was 90°. This was because when the inner fibre was wound at 90°, the entire FRP tube not only exerted a certain hoop restraint on the internal rebar but also provided a certain longitudinal tension, thereby reducing the generation of circumferential cracks and slowing down the occurrence of buckling. However, if it is hoop winding, the FRP hoop restraint is very strong. Once buckling occurs, multiple hoop cracks appear in the specimen, which accelerates the occurrence of buckling.

3.2.3. Analysis of Effect of Winding Thickness on FCCC-R Axial Compression Performance

From Table 5 and Figure 6b, it is apparent that when the fibre winding angle and slenderness ratio were constant, there was generally no difference in the first ascending branch between the FCCC-R specimens with different fibre winding thicknesses, but the length of the second ascending branch was different. The longer the fibre winding thickness, the longer the length of the second ascending branch. The data of Table 5 show that the peak loads of the FCCC-R specimens with 2, 4, and 6 fibre winding layers were 30%, 48%, and 68% higher than those of the corresponding pure steel bars, respectively. The axial displacements corresponding to the peak loads were 214%, 390%, and 576% higher than those of the corresponding pure steel bars, respectively.

Therefore, when the fibre winding angle and slenderness ratio were constant, the peak load and ductility of the FCCC-R specimens increased gradually with an increase in the FRP winding thickness. This was because with an increase in the FRP winding thickness, the restraint and longitudinal tension provided by the FRP increased, which limited the transverse deformation of the specimens [25] and thereby slowed down the occurrence of buckling.

3.2.4. Analysis of Effect of Slenderness Ratio on FCCC-R Axial Compression Performance

It is evident from Table 5 and Figure 6c that when the fibre winding angle and winding thickness were constant, FCCC-R specimens with different slenderness ratios differed in terms of the second ascending branch. The larger the slenderness ratio, the shorter the length of the second ascending branch, while the linear stage almost remained unchanged. The data of Table 5 show that the peak loads of the FCCC-R specimens with slenderness ratios of 22.73, 20, and 15.45 were 15%, 31%, and 53% higher than those of the corresponding pure steel bars, respectively. The axial displacements corresponding to the peak loads were 188%, 390%, and 760% higher than those of the corresponding pure steel bars, respectively. In addition, the slenderness ratio had a greater impact on the FCCC-R specimens than on the pure steel bars, as evident from Table 5, and it had a more significant effect on the FCCC-R axial compression performance than the FRP winding angle and thickness.

Therefore, when the fibre winding angle and winding thickness were constant, the peak load and ductility of the FCCC-R specimens decreased as the slenderness ratio increased. There were two reasons for this: first, with an increase in the slenderness ratio, the non-uniformity of the circumferential strain distribution of the FRP cloth in the height direction increased, and the lateral restraint effect decreased, which could not ensure the high performance of the axial compression of the specimens and hence led to a decrease in the bearing capacity of the specimens. Second, the lateral deflection of the specimens increased with the slenderness ratio. It is noted that for specimens with a slenderness ratio greater than 12.5, lateral deflection occurs at the initial loading stage, and it gradually changes from linear growth to nonlinear growth with an increase in the load. When the load is on the verge of reaching the ultimate load, the lateral deflection of the specimens increases sharply. Finally, instability and failure occur because of excessive lateral deflection [26]. Therefore, the greater the slenderness ratio of the specimen, the greater the lateral deflection, which accelerates the occurrence of buckling.

4. Finite Element Simulation and Parametric Analysis

4.1. Material Constitutive Model

For the steel rebars, the elastic–plastic trilinear constitutive model with a yield plateau was adopted, and for the mortar materials, we used the concrete damage plastic model in [4] and its compression and tension constitutive relations. As for the FRP materials, the mechanical behaviour of FRP was divided into elastic and damage stages, and lamina elastic material was used in the elastic stage [27–29]. The Hashin damage criterion was used to define the fracture failure of the composite materials in the damage stage [30–32]. Specific data are shown in Table 3.

4.2. Establishment of Finite Element Model

Since the FCCC-R specimens comprised an internal steel bar, mortar, and an FRP pipe, the solid element and shell element were used in ABAQUS to separately establish these three components, as shown in Figure 7a–c. Among them, the C3D8R solid element was used for the mortar and internal steel bar, and the S4R shell element was used for the FRP pipe because its thickness was very small compared with the cross section of FCCC-R. The three components were assembled and grids were divided to integrate the entire FCCC-R specimen, as shown in Figure 7d.

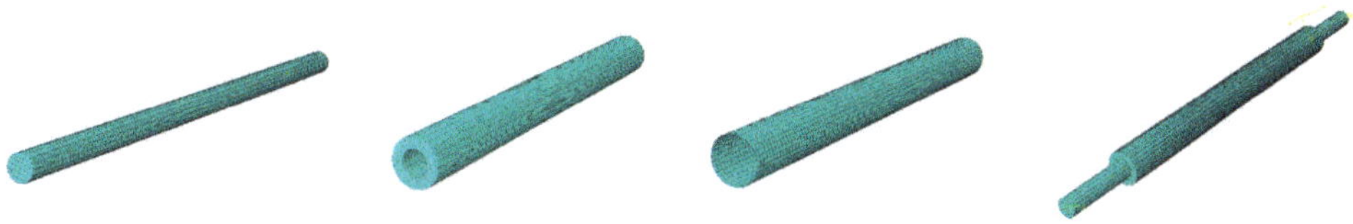

(**a**) Internal steel. (**b**) Mortar. (**c**) FRP pipe. (**d**) FCCC-R.

Figure 7. Configuration of FCCC-R model.

4.3. Verification Result of Finite Element

4.3.1. Failure Mode

As shown in the experimental results, there were two main failure modes of FCCC-R. First, owing to the circumferential winding of the inner layer fibre, the tension in the longitudinal direction was provided only by the fibre glue. However, the tensile strength of a colloid is low. Therefore, there were multiple circumferential cracks in the middle of the specimens, causing damage to the specimens. Second, the angled winding of the inner fibre led to the fibre providing a certain longitudinal tension, and the outer composite colloid was subjected to the longitudinal tensile action of the inner fibre. Therefore, the failure of the specimen was circumferential crack propagation. ABAQUS simulated the two failure modes well, and its simulations were consistent with the test results. The simulation and experimental failure modes of the F-400/440-4-a and F-400/440-4-c specimens are shown in Figure 8.

(**a**) F-400/440-4-a. (**b**) F-400/440-4-c.

Figure 8. Comparison between simulated and experimental failure modes of FCCC-R specimens.

4.3.2. Load–Displacement Curve

ABAQUS was used to simulate the seven groups of FCCC-R specimens under monotonic axial compression, and the load–displacement curves obtained are shown in Figure 9. Clearly, the match between simulation and experimental load–displacement curves is very good at the elastic stage, and the curves show the same trend after entering the elastic–plastic stage.

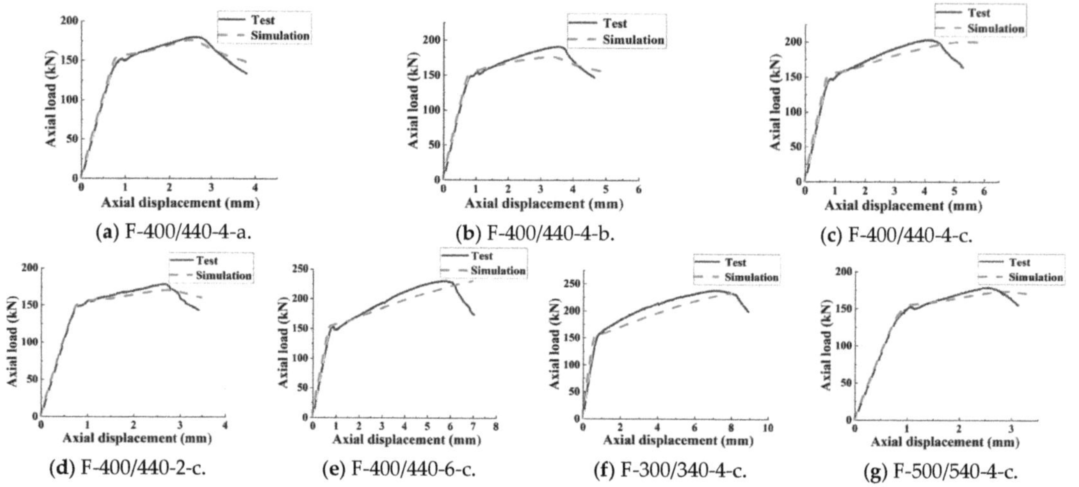

(a) F-400/440-4-a. (b) F-400/440-4-b. (c) F-400/440-4-c. (d) F-400/440-2-c. (e) F-400/440-6-c. (f) F-300/340-4-c. (g) F-500/540-4-c.

Figure 9. Load–displacement curve comparison for FCCC-R specimens.

The test and simulated values of the peak load for each specimen are listed in Table 6. $P_{\max}$ is the peak load measured in the test, and $P_{\max,M}$ is the peak load calculated by the simulation. It can be seen from the results that the maximum error between the two values is 7.30%, and the average ratio is 1.03. The simulation results are in good agreement with the test results, indicating that the finite element model can simulate the load–displacement relationship of the FCCC-R specimens under monotonic axial compression accurately.

Table 6. Comparison between the test value and the simulation value of peak loads.

Specimen ID	P_{max} (kN)	$P_{max,M}$ (kN)	$P_{max}/P_{max,M}$	Discrepancy
F-400/440-4-a	180.16	176.00	1.02	2.31%
F-400/440-4-b	190.90	176.97	1.08	7.30%
F-400/440-4-c	203.69	200.84	1.01	1.40%
F-400/440-2-c	178.83	170.64	1.05	4.58%
F-400/440-6-c	231.00	230.78	1.00	0.10%
F-300/340-4-c	238.17	233.48	1.02	1.97%
F-500/540-4-c	178.55	173.98	1.03	2.56%

4.4. *Parametric Study and Analysis*

In order to further study the factors affecting the axial compression performance of FCCC-R and make up for the limitations of the objective conditions in the test, on the basis of the finite element model constructed, the influence parameters that could not be studied in the test were analysed. The basic dimensions of the finite element model were determined based on the F-400/440-4-a group. The parameters studied in this section were the area ratio A/A_s, confinement length L_r/L, internal bar position eccentricity e/R, and mortar strength f_c. Table 7 lists the ranges of all parameters considered in the parametric study.

Table 7. Values of the parameters considered in the parametric study.

Parameter	Values
Area ratio, A/A_s	2.53, 4.18, 6.25, 10.12, 16.74, 25
Confinement length, L_r/L	0.52, 0.60, 0.69, 0.79, 0.90
Internal bar position eccentricity, e/R	0.6, 0.45, 0.30, 0.15, 0
Mortar strength, f_c (MPa)	20, 40, 60

The area ratio A/A_s of FCCC-R is defined as the ratio of the FRP pipe's internal area A to the internal bar's cross-sectional area A_s. As shown in Figure 10a, the peak stress and average strain increased dramatically with an increase in the area ratio. For a small area ratio, such as the area ratio of 2.53, there was no second ascending branch, and the axial load decreased immediately after the peak load. This was because the smaller the area ratio, the smaller the moment of inertia of the section provided, which was insufficient to resist the influence of the additional moment caused by the lateral displacement. The additional moment in the middle of the span of the specimen was too large, and hence the rebar buckled, causing the premature failure of the specimen. Thus, a large area ratio is preferable in the design of the FCCC-R specimen.

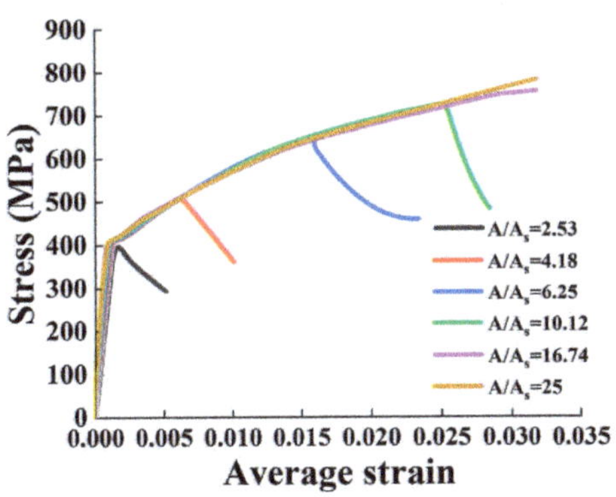

(**a**) Area ratio.

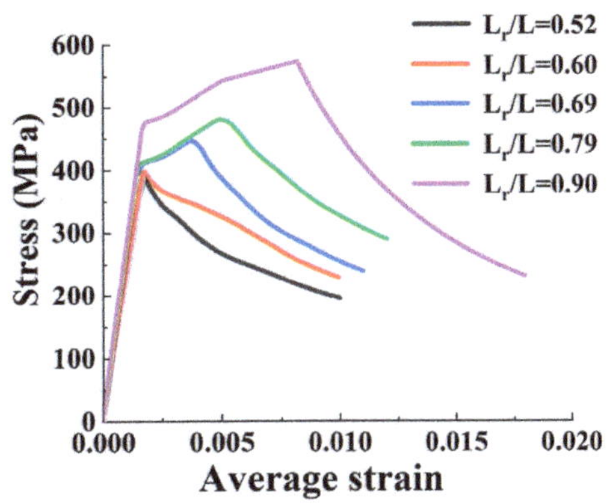

(**b**) Confinement length.

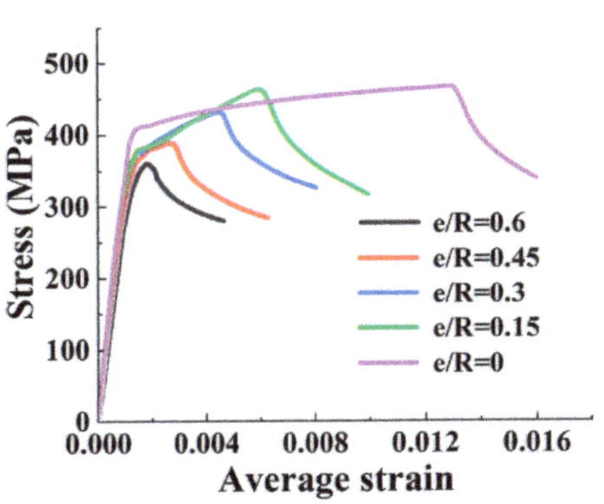

(**c**) Internal bar position eccentricity.

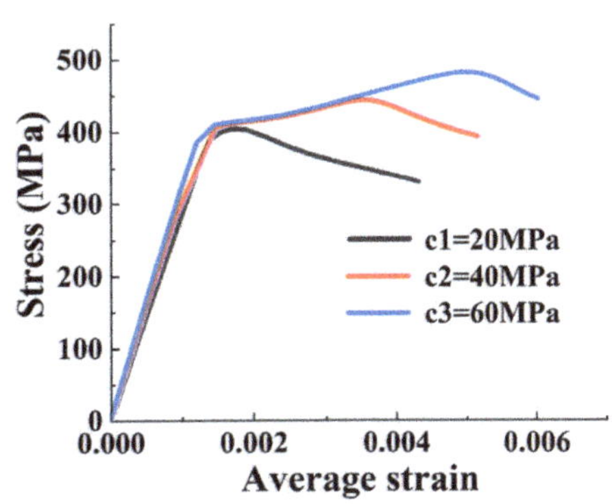

(**d**) Mortar strength.

Figure 10. Effect of different parameters on the average stress–strain relationships under compression.

FCCC-R's confinement length ratio L_r/L is defined as the ratio of the FRP pipe's length L_r to the internal bar's length L. If FCCC-R is placed in a concrete column, the steel bar may not be entirely confined by the FRP pipe because of the consideration of the steel bar anchorage and connection. The two ends of the steel bar are intended to remain outside the FRP pipe. Hence, it is necessary to investigate the influence of confinement length. As shown in Figure 10b, the specimens' bearing capacity and ductility improved to a certain extent with an increase in the confinement length, and the greater the confinement length ratio, the more obvious the improvement was. This was because a longer FRP confinement length resulted in a stronger constraint action and higher longitudinal tension on the specimen, which resulted in poorer circumferential deformation and slowed down the occurrence of buckling. Thus, a large restrained length ratio is preferable in the design of FCCC-R specimens.

The internal bar position eccentricity e/R of FCCC-R is defined as the ratio of the distance e from the centre of the inner reinforcement to the centre of the FRP pipe to the inner radius R of the FRP pipe. It can be found from Figure 10c that the specimens' axial compressive behaviour gradually deteriorated and their bearing capacity along with ductility continued to decrease as the bar's eccentricity increased. This can be explained as follows: with an increase in the internal bar eccentricity, the restraint effect of FRP and concrete on one side of the reinforcement is smaller than that at the central position, which accelerates the occurrence of buckling. Therefore, a better configuration of a single FCCC-R can be obtained by placing the rebar at the centre of the cross section under axial compressive loads.

Figure 10d shows the effect of the mortar strength on FCCC-R's axial compression performance. The peak load slightly increases as the mortar strength increases, and its influence is much smaller compared with that of other parameters. However, the behaviour after the peak is almost the same. The reason was that with the mortar cracking under tension, all specimens tended to fail without the mortar being crushed. Thus, the mortar provided only tensile strength, and the tensile strength of the mortar was very low. Therefore, the use of mortars with different strengths had little impact on the axial compression performance of FCCC-R.

5. Conclusions

In this study, the buckling performance of FCCC-R under monotonic axial compression was analysed by performing the monotonic buckling test and a simulation on commercial finite element software. The main conclusions are as follows:

(i) The axial compression performance of FCCC-R was found to be superior to that of pure steel bars due to the dual constraints of FRP and concrete. The load–displacement curve of FCCC-R demonstrated a second ascending branch, characterised by high bearing capacity and ductility. As a result, the internal steel bars with a high slenderness ratio were able to reach or surpass the yield strength during buckling, effectively avoiding premature buckling under compression.

(ii) The experimental results indicated that the most optimal fibre winding angle for FCCC-R was 90°. The axial compression performance of FCCC-R was observed to improve with an increase in the thickness of FRP winding. Moreover, the slenderness ratio was identified as the primary factor affecting the axial compression performance of FCCC-R. A reduction in the slenderness ratio resulted in a noticeable increase in the length of the second ascending branch in the load–displacement curve. To be specific, a decrease in the slenderness ratio from 22.73 to 15.45 resulted in an increase in the bearing capacity and ductility of the specimens by 33% and 175%, respectively.

(iii) A finite element model of FCCC-R was established using the ABAQUS software. The model was validated through comparison with experimental results, with a maximum discrepancy of 7.30%. The simulation results were found to be in accordance with the experimental outcomes, indicating the reliability and validity of the established model.

(iv) A parametric analysis was conducted, revealing that the area ratio is the most critical factor that affects the axial compression performance of FCCC-R. As such, it is advisable to increase the area ratio in the design of FCCC-R. Moreover, improvements in the axial compression behaviour of FCCC-R can be achieved by increasing the FRP confinement length and reducing the eccentricity of reinforcement. In particular, it was observed that the mortar strength had a minimal impact on the axial compression performance of FCCC-R.

Author Contributions: Conceptualization, J.L.; methodology, J.L.; software, H.H.; validation, H.H.; formal analysis, H.H.; investigation, T.M.; resources, J.L.; data curation, Y.L.; writing—original draft preparation, H.H.; writing—review and editing, J.L.; visualization, H.H.; supervision, J.L.; project administration, J.L.; funding acquisition, J.L. All authors have read and agreed to the published version of the manuscript.

Funding: This research was funded by the National Natural Science Foundation of China (Grant No. 51978596).

Institutional Review Board Statement: Not applicable.

Informed Consent Statement: Not applicable.

Data Availability Statement: Data sharing is not applicable to this article.

Conflicts of Interest: The authors declare no conflict of interest.

References

1. Syntzirma, D.V.; Pantazopoulou, S.J.; Aschheim, M. Load-history effects on deformation capacity of flexural members limited by bar buckling. *Eng. Struct.* **2009**, *136*, 1–11. [CrossRef]
2. Massone, L.M. Fundamental principles of the reinforced concrete design code changes in chile following the mw 8.8 earthquake in 2010. *Eng. Struct.* **2013**, *56*, 1335–1345. [CrossRef]
3. Kashani, M.M.; Lowes, L.N.; Crewe, A.J.; Alexander, N.A. Nonlinear fiber element modelling of RC bridge piers considering inelastic buckling of reinforcement. *Eng. Struct.* **2016**, *116*, 163–177. [CrossRef]
4. Wang, Z.Y.; Feng, P.; Zhao, Y.; Yu, T. FRP-confined concrete core-encased rebar for RC columns: Concept and axial compressive behavior. *Compos. Struct.* **2019**, *222*, 110915.1–110915.13. [CrossRef]
5. Dhakal, R.P.; Su, J. Design of transverse reinforcement to avoid premature buckling of main bars. *Earthq. Eng. Struct. Dyn.* **2018**, *47*, 147–168. [CrossRef]
6. Hashemi, S.S.; Vaghefi, M.; Hemmat, M. Evaluation the effects of stirrup spacing and buckling of steel reinforcing bars on the capacity of RC columns. *Sci. Iran.* **2017**, *25*, 1140–1151. [CrossRef]
7. Pereiro-Barceló, J.; Bonet Senach, J.L.; Albiol-Ibáñez, J.R. Required tie spacing to prevent inelastic local buckling of longitudinal reinforcements in RC and FRC elements. *Eng. Struct.* **2018**, *160*, 328–341. [CrossRef]
8. Lukkunaprasit, P.; Tangbunchoo, T.; Rodsin, K. Enhancement of seismic performance of reinforced concrete columns with buckling-restrained reinforcement. *Eng. Struct.* **2011**, *33*, 3311–3316. [CrossRef]
9. Mitra, D.C.; Bindhu, K.R. Seismic performance of RC short columns with buckling restraint reinforcement. *Mag. Concr. Res.* **2019**, *71*, 163–174. [CrossRef]
10. Ruangrassamee, A.; Sawaroj, A. Seismic enhancement of reinforced-concrete columns by rebar-restraining collars. *J. Earthq. Tsunami.* **2013**, *6*, 1250015.1–1250015.19. [CrossRef]
11. Tang, Y.; Lu, X.; Wei, Y.; Hou, S. Experimental study on compressive behavior of concrete cylinders confined by a novel hybrid fiber-reinforced polymer spiral. *Polymers* **2022**, *14*, 4750. [CrossRef] [PubMed]
12. Ye, Y.Y.; Zeng, J.J.; Li, P.L. A state-of-the-art review of FRP-confined steel-reinforced concrete (FCSRC) structural members. *Polymers* **2022**, *14*, 677. [CrossRef]
13. Zhou, Y.; Hu, J.; Li, M.; Sui, L.; Xing, F. FRP-confined recycled coarse aggregate concrete: Experimental investigation and model comparison. *Polymers* **2016**, *8*, 375. [CrossRef]
14. Wu, G.; Lu, Z. Study on the stress-strain relationship of FRP-confined concrete rectangular columns. *J. Build. Struct.* **2004**, *25*, 99–106.
15. Li, P.; Wu, Y.F. Stress–strain model of FRP confined concrete under cyclic loading. *Compos. Struct.* **2015**, *134*, 60–71. [CrossRef]
16. Micelli, F.; Mazzotta, R.; Leone, M.; Aiello, M.A. Review study on the durability of FRP-confined concrete. *J. Compos. Construct.* **2015**, *19*, 04014056. [CrossRef]
17. Li, P.; Huang, D.; Li, R.; Yuan, F. Effect of aggregate size on the axial compressive behavior of FRP-confined coral aggregate concrete. *Polymers* **2022**, *14*, 3877. [CrossRef]
18. Raza, A.; Ahmad, A. Prediction of axial compressive strength for FRP-confined concrete compression members. *Civ. Eng.* **2020**, *24*, 2099–2109. [CrossRef]

19. Teng, J.G.; Lam, L. Compressive behavior of carbon fiber reinforced polymer-confined concrete in elliptical columns. *J. Struct. Eng.* **2002**, *128*, 1535–1543. [CrossRef]
20. Cao, Y.; Wu, Y.F.; Jiang, C. Stress-strain relationship of FRP confined concrete columns under combined axial load and bending moment. *Compos. Part B Eng.* **2018**, *134*, 207–217. [CrossRef]
21. Yu, T.; Lin, G.; Zhang, S.S. Compressive behavior of FRP-confined concrete-encased steel columns. *Compos. Struct.* **2016**, *154*, 493–506. [CrossRef]
22. Feng, P.; Zhang, Y.; Bai, Y.; Ye, L. Strengthening of steel members in compression by mortar-filled FRP tubes. *Thin-Walled Struct.* **2013**, *64*, 1–12. [CrossRef]
23. Jiang, C. Strength enhancement due to FRP confinement for coarse aggregate-free concretes. *Eng. Struct.* **2023**, *277*, 115370. [CrossRef]
24. Zhao, D.B.; Pan, J.; Zhou, Y.W.; Sui, L.L.; Ye, Z.H. New types of steel-FRP composite bar with round steel bar inner core: Mechanical properties and bonding performances in concrete. *Constr. Build. Mater.* **2020**, *242*, 118062. [CrossRef]
25. Deng, Z.C.; Liu, S.X. Experimental and theoretical research on ultra-high performance concrete confined by FRP pipe. *J. Basic Sci. Eng.* **2016**, *24*, 792–803.
26. Hu, Z.J.; Pan, J.L.; Xu, T. Experimental research on stable carrying capacity of FRP strongly confined concrete reinforced columns. *Ind. Arch.* **2005**, *35*, 24–28.
27. Cang, Y.Q. Research on Axial Compression and Seismic Performance of Steel Tube-GFRP-Sea Sand Concrete Column Based on Unified Theory. Master's Thesis, Harbin Institute of Technology, Harbin, China, 2010.
28. Liu, L.J. Experimental and Theoretical Study on Stability of Pultrusion GFRP Axial Compressive Members. Master's Thesis, Nanjing University of Science and Technology, Nanjing, China, 2020.
29. Ren, H.T.; Guo, X.; Wang, S.Y. Axial compression performance test and finite element analysis of FRP pipe-concrete-steel tube composite short columns. *J. Dalian Univ. Technol.* **2015**, *55*, 612–617.
30. Chaht, F.L.; Mokhtari, M.; Benzaama, H. Using a Hashin criteria to predict the damage of composite notched plate under traction and torsion behavior. *Frat. Integrita Strutt.* **2019**, *50*, 331–341. [CrossRef]
31. Son, J.; Fam, A. Finite element modeling of hollow and concrete-filled fiber composite tubes in flexure: Model development. *Verif. Investig. Tube Parameters Eng. Struct.* **2008**, *30*, 2656–2666.
32. Liu, W.L.; Chang, X.L.; Zhang, X.J.; Hu, K.; Zhang, Y.H. Research on low velocity impact damage of composite materials based on improved Hashin criterion. *Vib. Impact* **2016**, *35*, 209–214.

polymers

MDPI

Article

Experimental and Numerical Investigation of Joints for a Pultruded Fiber-Reinforced Polymer Truss

Yiwei Chen [1], Maojun Duan [1], Xingxing Zou [1,*], Yu Feng [2] and Guofen Li [1]

[1] College of Civil Engineering, Nanjing Forestry University, Nanjing 210037, China
[2] Shanghai Construction (Group) Co., Ltd., Shanghai 200080, China
* Correspondence: civilzou@njfu.edu.cn; Tel.: +86-15586188579

Abstract: Bolted connections usually govern the structural rigidity and load-carrying capacity of pultruded glass fiber-reinforced polymer (GFRP) truss structures. In this study, a novel bolted integrated gusset plate (IGP) connection is proposed to enhance the stiffness and capacity of GFRP truss structures. Nine double-lap shear tests of GFRP joints and numerical simulation were conducted to investigate the influence of variable design parameters of the bolted GFRP joints (number of bolts, width and thickness of GFRP, edge distance of bolts, and the employment of adhesive). Three full-scale GFRP truss joints were tested under static loading to study the response of a typical bolted connection, a bolted gusset plate connection, and the proposed IGP connection. The nine double-lap shear tests showed that the bolted–bonded mixed connection has 50% higher shear stiffness and 27% higher ductility compared with bolted joints, and bearing failure dominated the capacity of most specimens, which agreed well with numerical simulation results. Tests on the three full-scale GFRP truss joints showed that the bolted gusset plate can substantially reduce the number of cracks and improve the initial stiffness, but the maximum bearing capacity of the joints did not increase because the shear fracture of pultruded GFRP webs governs the capacity. The proposed IGP substantially increased the stiffness and capacity compared with the bolted connection and typical bolted gusset plate connection. The full-scale GFRP joint test is suggested to be used together with direct shear tests to study the performance of joints of the GFRP truss.

Keywords: pultruded GFRP truss; bolted connection; bolted-bonded connection; integrated gusset plate; shear performance

1. Introduction

Fiber-reinforced polymers (FRPs) have been made into various products and used in different civil engineering structures because of their advantages of light weight, high strength, and excellent corrosion resistance [1,2]. Different fiber types of FRP products result in different mechanical properties and corrosion, fatigue, creep resistances [3]. Typical FRPs can be divided into three main categories by fiber type: glass-, carbon- and basalt-fiber-reinforced polymers (GFRP, CFRP, and BFRP). Among them, GFRP is most widely used for its relatively low cost [4]. CFRP has outstanding strength, modulus and corrosion resistance, although the cost is relatively high. BFRP has good insulation, corrosion resistance, high temperature resistance, and recyclability [5]. Among various manufactory techniques (hand layup, pultrusion, resin transfer molding, filament wound, etc.) of FRP products, pultruded GFRP profiles have been increasingly used in civil engineering structures such as railway sleepers, hollow slabs, truss bridges, and space-formed buildings in recent years [6–9].

GFRP truss structures demonstrate light weight, fast construction, and extraordinary durability. Joints that connect different GFRP truss members play a critical role in transferring loads and ensuring structural integrity [10]. In field applications of truss bridges and buildings with pultruded GFRP profiles, joints are typically connected by bolted or bolted-bonded mixed connection [8]. Previous studies have revealed that GFRP joints usually govern the capacity and rigidity of the overall structure [11–14]. At present, the open and closed cross sectional profile clamping bolted connection is widely used in GFRP pultruded profile truss structures. Owing to the low shear strength of pultruded GFRP profiles, GFRP trusses are prone to brittle failure modes of web shear cracks along the profile axis under the action of bending moment within the joints [15,16]. Feng et al. conducted a full-scale test of GFRP truss joints, which revealed that the truss with GFRP pultruded profiles is a structural system suitable for bridges because of its high stiffness, bearing capacity and load efficiency, but the joints need to be further studied [8]. Keller et al. conducted experimental research on the two-span Pontresina truss bridge and showed that the span with a bolted–bonded mixed connection resulted in higher stiffness of the overall GFRP truss bridge than the bolted span at both short and long serviceability [17,18]. It is widely known that the joints of a GFRP truss bridge restrict the development of the truss bridge [19–22], and so proposing GFRP joints with a high capacity and stiffness is of pivotal importance.

In order to research the performance of GFRP joints, tests can be divided into two categories: (i) single- or double-lap direct shear tests of GFRP plates, and (ii) loading tests on full- or large-scale GFRP joints [23,24]. Direct shear test results showed that a bolted–bonded mixed connection has less deformation than bolted or bonded connections, but their ultimate capacity is very similar because they have the same failure modes [25,26]. Four typical failure modes were observed in the tests of bolted pultruded GFRP profiles: bolt bearing failure, GFRP plate tensile failure, GFRP plate shear failure, and GFRP plate splitting failure [27–29]. The load capacity and failure mode of the GFRP plate are related to the geometric dimensions such as hole diameter (φ), width of the GFRP plate (w), end distance of the GFRP plate (e) and fiber layup of the GFRP plate [30–33].

Compared with the direct shear test, the experimental research on full- or large-scale GFRP joints is scare. Chen carried out an experimental test and numerical analysis on a GFRP truss bridge where gusset plates were used in the joints [34]. The results showed that the gusset plate reduced the maximum stress occurring in the middle of the bottom surface of the lower chord of the midspan section. The stress value was far lower than the material strength, which can meet the needs of practical use. Ascione et al. [35] and Russo [36,37] proposed the use of pultruded profile gusset plates to connect more chords, but the end shear splitting damage of profile gusset plates was more severe under loading. In summary, the failure of a pultruded GFRP truss bridge mainly depends on the connection strength at the joint because of the local stress concentration around the connection region, but the research on full-scale pultruded GFRP profiles joint is also not extensive enough at present [38–41]. Whether the simple direct shear test can accurately reflect the real joints' loading transferring paths in field applications of GFRP trusses and how it is related to the full- or large-scale GFRP joints' test are still not clear currently [42–44]. Additionally, how the gusset plates change the load response of the GFRP joints is not clear at present [45,46].

In view of the above problems, a novel bolted integrated gusset plate connection is proposed and is expected to achieve better stiffness and capacity than typical a bolted connection and a bolted connection with a gusset plate. Considering the lack of research on bolted–bonded hybrid connections, this study presents an experimental and numerical investigation of a double-lap direct shear test. Full-scale testing of the proposed integrated gusset plate connection was conducted. The relationship between the double-lap direct shear test and the full-scale test of the connection is analyzed.

2. Materials and Methods

2.1. Materials

The materials of the GFRP pultrusion profiles used in the tests (Figure 1a) are commercially available products from Nanjing Kangte composite materials company in Nanjing China. The resin used is unsaturated resin and the fiber is glass fiber. The material properties of the specimens were obtained directly from the walls of the pultruded profiles by an automatic CNC engraving machine (Tianxiang Inc., Tengzhou, China). Five tensile tests along and perpendicular to the fiber direction (Figure 2a), five compression tests (Figure 2b) and five longitudinal interlaminar shear tests (Figure 2c) were completed, and the results used were the average of the tests. The tensile strength of the GFRP profile was 308 MPa, the compressive strength was 210 MPa and the interlayer shear strength was 18.2 MPa, which were experimentally obtained according to Chinese code [47]. The modulus of elasticity of GFRP was 46 GPa according to the manufacturer. The bolts (see Figure 1b) used in the test were external hexagonal grade 8.8 high-strength M12 bolts, with a tensile strength of 800 MPa and yield strength of 640 MPa [12]. The modulus of elasticity of steel bolts was 210 GPa according to the manufacturer. The structural adhesive was an epoxy resin adhesive with shear strength $\geq$ 12 MPa and temperature resistance of -60 °C to 120 °C according to the manufacturer.

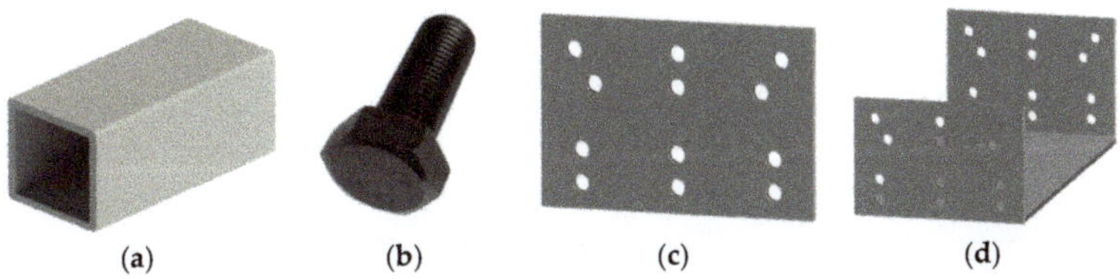

Figure 1. Materials used in tests of this study: (**a**) pultruded GFRP profiles, (**b**) steel bolt, (**c**) gusset plate, and (**d**) proposed integrated gusset plate.

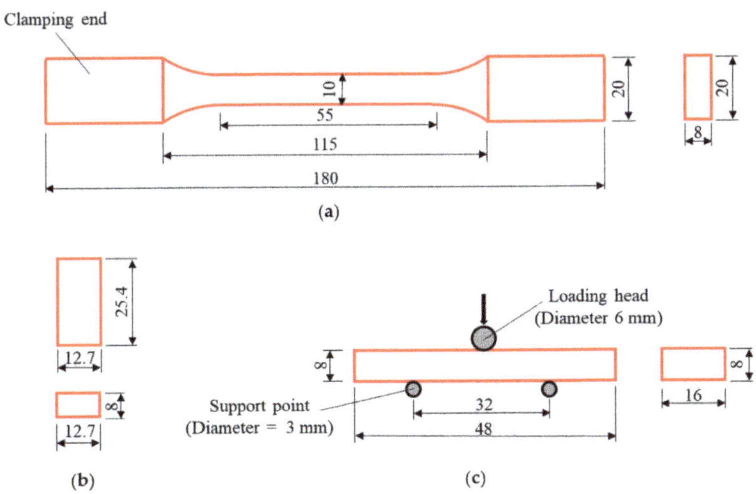

Figure 2. Geometric sizes of material test coupons: (**a**) tensile test, (**b**) compression test, and (**c**) shear test (units in mm).

The GFRP and high-strength bolts used in the three GFRP truss tests were the same as in the double-lap direct shear tests. The Q235 steel gusset plate (see Figure 1c,d) was used to combine different members in the truss series, with a density of 7850 kg/m^3 and a modulus of elasticity of 200 GPa.

2.2. Methods

2.2.1. Double-Lap Direct Shear Test of Bolted GFRP Plates

Specimen Design of Double-Lap Direct Shear Test

Referring to the experiments of Russo et al. [19,33], this study selected the number of bolts, plate width, plate end distance, plate thickness, and adhesive as variables and designed nine groups of GFRP profile bolted connection specimens. Test parameters and specimen configurations are shown in Table 1. Among them, the T1 specimen was a thick plate standard piece, and the changes in the T2–T9 specimens were as follows: T2—reduced the plate width w, T3—reduced the end distance e, T4—increased e, T5—added adhesive at the interface, T6—reduced the plate thickness t, T7—reduced t and used a single bolt, T8—reduced t and used three bolts, T9—reduced t and added adhesive at the interface.

Table 1. Summary of the main parameters of GFRP bolted connection specimens.

Specimen ID	l (mm)	w (mm)	t (mm)	p (mm)	e (mm)	φ (mm)	P_u (kN)	P_{bolt} (kN)
T1	350	60	10	70	50	12	100.7	50.4
T2	350	45	10	70	50	12	80.8	40.4
T3	350	60	10	70	40	12	89.3	44.7
T4	350	60	10	70	60	12	94.5	47.3
T5	350	60	10	70	50	12	97.9	48.9
T6	350	60	6.5	70	50	12	62.1	31.0
T7	350	60	6.5	70	50	12	35.2	35.2
T8	420	60	6.5	70	50	12	65.5	21.8
T9	350	60	6.5	70	50	12	63.2	31.6

Construction of Double-Lap Direct Shear Test Specimens

Double-lap direct shear specimens were obtained directly from the flanges and webs of the pultruded profiles by an automatic CNC engraving machine (Tianxiang Inc., Tengzhou, China). Then, the locations of bolts were marked, and the drilling was completed (as shown in Figure 3). The bolts were assembled by a trench that can measure the torque which is related to the pre-tightening stress within the bolts. For bolted mixed GFRP members, before tightening the bolts, the overlapping part was sanded and then evenly coated with epoxy resin adhesive to ensure sufficient contact between the adhesive and the surface.

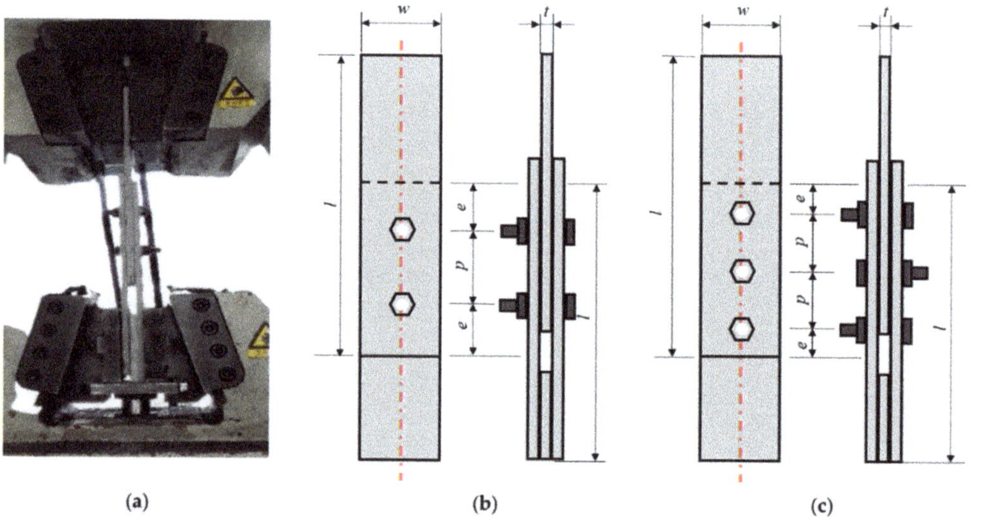

Figure 3. Double-lap shear test setup and geometric configuration of specimens: (**a**) actual photo, (**b**) specimen with two bolts, and (**c**) specimen with three bolts.

Data Acquisition of Double-Lap Direct Shear Test

The GFRP double-lap direct shear connector test (Figure 3) used the SANS-SHT4106 universal testing machine (Sansi Eternal Technology Co., Ningbo, China). In order to eliminate the possible gap between the bolt and the plate, it was first loaded at a speed of 2 mm/min to reach 20 kN and then unloaded, and finally the main loading started at a loading speed of 2 mm/min. The test was terminated once the part was damaged or the specimen experienced a large displacement.

2.2.2. Full-Scale Test of GFRP Pultruded Profile Truss Joints

Specimen Design of Full-Scale Test of GFRP Pultruded Profile Truss Joints

One full-scale typical clamping truss joint (specimen BJ, which stands for bolted joints), one typical gusset plate connection joint (specimen GP, which stands for bolted gusset plate joint), and one proposed integrated gusset plate connection joint (specimen IGP, which stands for integrated gusset plate joint) were designed. Each specimen was composed of a single 1000 mm lower chord web chord, two 450 mm 45° diagonal chords, a single 450 mm vertical chord, and a steel gusset plate connection. The cross-sectional dimensions of the pultruded section bars used for the joint specimens were the same, all of which were 100 mm × 100 mm closed rectangular sections. In order to maintain the consistency in the number of pultruded profiles, the required groove chords were cut by a CNC machine along the axis of rectangular section square tubes. Among them, the web of bottom chords of the typical clamping truss joint consisted of two 1000 mm web chords with a sectional size of 50 mm × 100 mm. The steel gusset plate connection-type truss joint was composed of two 1000 mm profile channel profiles, the lower chord web chord, and four 450 mm profile channel profiles to form a 45° diagonal chord, and two 450 mm profile channel profiles to form a vertical web member. The sectional size of all profile channel profiles was 50 mm × 100 mm. The test setup can be seen in Figure 4a. The three GFRP pultruded profile truss joints are shown in Figure 4c–e.

Data Acquisition of Full-Scale Test of GFRP Pultruded Profile Truss Joints

A hydraulic servo actuator with a maximum range of 300 kN acted directly on the top of the vertical bar for loading. The displacement and strain data of the joint static performance test were recorded by a TDS-7130 high-speed static strain data acquisition system produced by Tokyo Institute of Instrumentation, Japan. Due to the rotation of the spherical joint at the loading end of the hydraulic servo actuator, the load eccentricity occurred in the test process. Before loading, the sub-solid high-precision horizontal laser instrument (accuracy error 1 mm within 5 m) was used to accurately locate the vertical axis of the actuator and the vertical chord of the specimen to ensure that the actuator was consistent with the vertical axis of the vertical chord. YHD-50 displacement measurement range is 50 mm, and the accuracy error is controlled within 0.05%.

During the loading process of the monotonic static test, the data of displacement and strain were collected. The deformation and load of the specimen under cracking loads were observed and recorded. The test loading mechanism was as follows: (i) we preloaded 2 kN to the specimen to inspect the setup and data acquisition system and then unloaded after ensuring that all instruments work normally; (ii) in the load test stage, a 2 mm/min displacement control mode was adopted for loading, and the structural response was monitored and saved in real time; (iii) after the specimen reached the ultimate load and failed, the specimen was unloaded gradually until the end of the test.

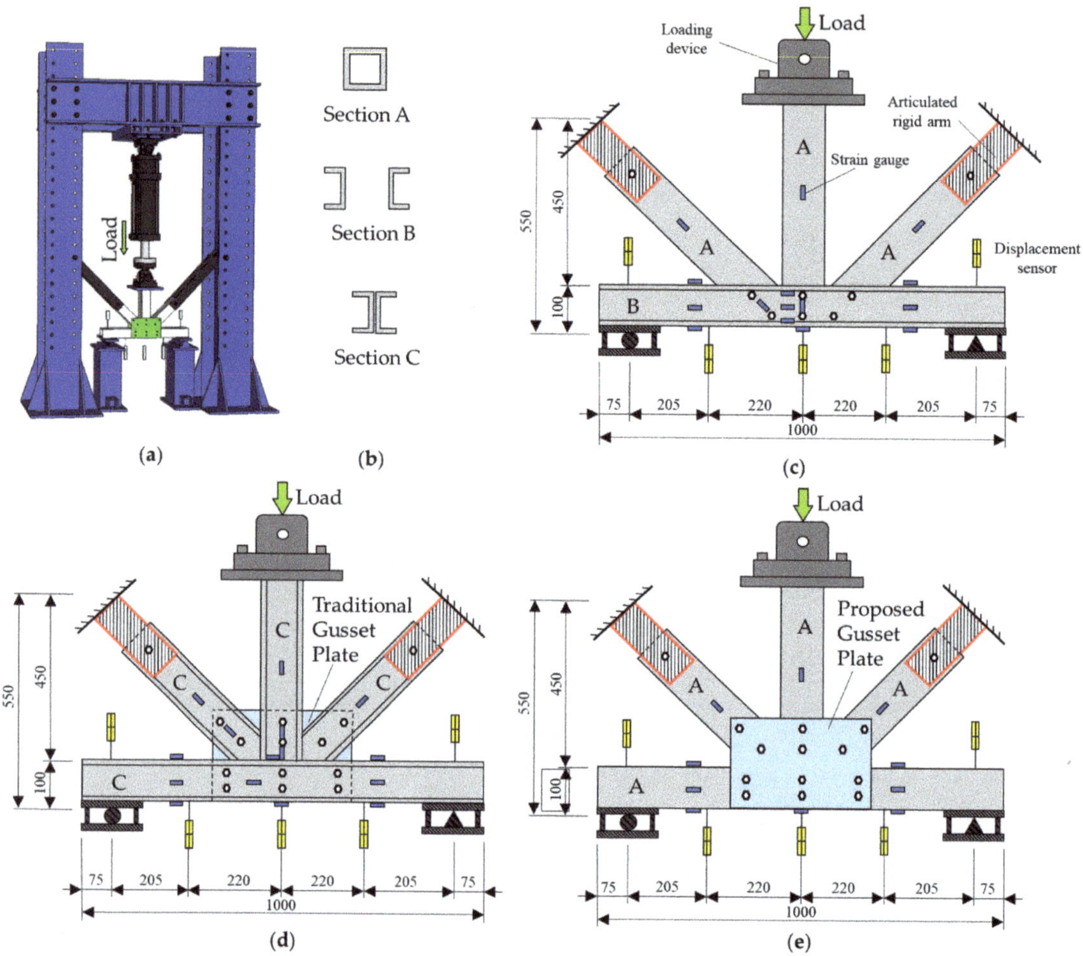

Figure 4. Full-scale test of GFRP truss joint: (**a**) test setup, (**b**) cross sections of members, (**c**) specimen BJ, (**d**) specimen GP, and (**e**) specimen IGP.

3. Results and Analysis of Double-Lap Shear Test

3.1. Failure Mode and Load Response of Double-Lap Direct Shear Test

Three failure modes formed in the double-lap shear test, as shown in Table 2. It can be found that when the w/φ is small, the GFRP tensile specimen is prone to tensile failure. When the w/φ is large, the GFRP plate is prone to bearing failure along the adjacent region of bolt holes. Between the above two cases, shear failure occurred along the pultrusion direction. The observed results, without considering e/φ, are consistent with the situation of the test results of Okutan [33] and also prove the reliability of the test results and showed that most double-lap shear tests failed in bearing failure. The development direction of cracks grew about 45° along the bolt hole perimeter, as shown in Figure 5.

Table 2. Classification of failure modes of GFRP bolted joints.

Failure Mode	Specimen	Typical Damage Diagram
Bearing	T1, T3, T4, T5, T7	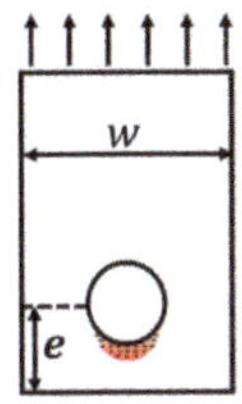
Tensile failure	T6, T9	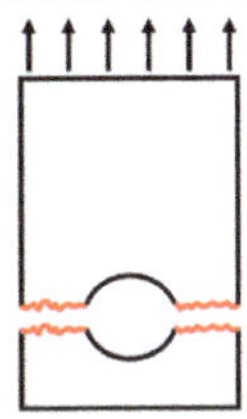
Shear-out	T2, T7, T8	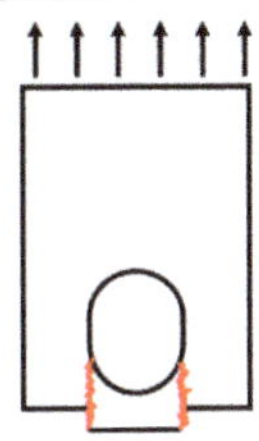

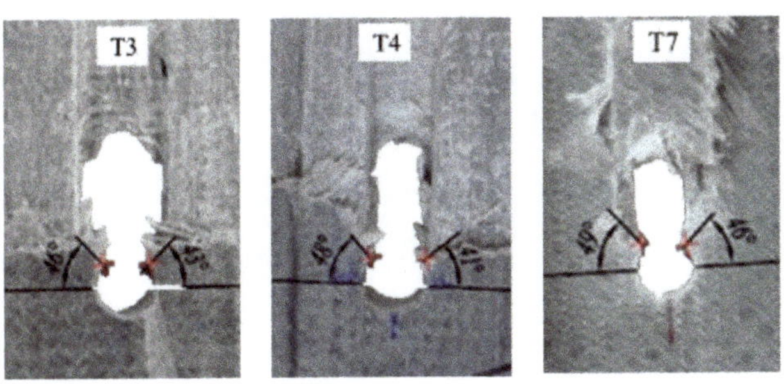

Figure 5. Shear failure at 45° of the bolt hole.

3.2. Parametric Study of Double-Lap Direct Shear Test

3.2.1. Number of Bolts

With the increase in the number of bolts, the bearing capacity of the connector increased, but the effect of increased load-bearing capacity does not show a multiplicative relationship. Before the T6 specimen reached the peak load, the load continued to rise linearly and the stiffness was basically unchanged; at 15 kN (0.24 P_u), due to the existence of bolt holes, the specimen experienced a small sudden drop in load, and the load returned to a linear rise quickly. When the peak load of the specimen was 62.1 kN (P_u), the load of the specimen decreased in a step shape. There were two descending plateaus, and the specimen showed good ductility characteristics.

In the early stage of the loading of the T7 specimen, the load basically increased linearly; at 8 kN (0.23 P_u) and 13 kN (0.37 P_u), due to the existence of bolt holes, the specimen had a small sudden drop in load, and the load returned to a linear rise quickly. When the load was continued, the specimen continued to make a fiber tearing sound. When the load was 31 kN (0.88 P_u), the tensile stiffness of the specimen decreased. When the peak load of the specimen reached 35.2 kN, the load of the specimen suddenly dropped to 21 kN (0.60 P_u), and there were three descending plateaus in total.

During the loading process of the T8 specimen, before 24 kN (0.37 P_u), the load basically continued to rise linearly, and there was a sound of a fine fiber rupture during the loading process. Between 24 kN and 33 kN (0.50 P_u), the stiffness of the specimen decreased due to the existence of bolt holes. After 33 kN, the stiffness of the specimen was restored. With the increasing of load, the specimen continued to make a fiber tearing sound until the peak load, and the specimen load maintained a linear increase. After reaching the peak load of 65.5 kN (P_u), the load of the specimen dropped suddenly and the displacement increased by 2 mm.

It can be seen in Figure 6 that when the number of bolts increased from 1 to 2, the bearing capacity of a single bolt (P_u) increased significantly, and the double-lap direct shear thin plate connector increased by nearly 76.3%. At the same time, as shown in Table 2, T6 experiences tensile failure, while T7 experiences shear-out failure, indicating that shear-out failure can achieve a higher bearing capacity, which is the failure mode we prefer to obtain. When raising from a double bolt to three bolts, the double-lap direct shear bolted gusset plate connection P_u only increased by 5.6%. It can be found from Figure 6 that the stiffness of the specimen was also increased with the increase in the number of bolts. Similarly, when the specimen was increased from a single bolt to a double bolt, the increase in the moving stiffness of the specimen was more obvious. The greater the number of bolts, the smaller the corresponding displacement when the specimen reached the ultimate load.

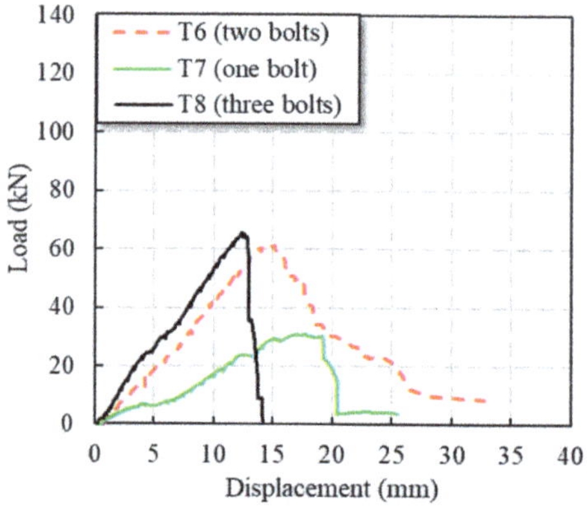

Figure 6. Load–displacement curves of specimens with different numbers of bolts.

3.2.2. Width of Plate

As Figure 7 shows, the T1 specimen basically maintained linear loading during the whole loading process. At 16 kN (0.16 P_u), there was a small sudden drop in load due to the existence of bolt holes. During the loading process after this, the specimen continuously emitted the sound of fibers tearing, which increased sharply in the later stage of loading. When the load reached the peak load of 100.7 kN (P_u), the specimen suddenly failed.

In the early stage of loading, the load of the T2 specimen basically continued to rise linearly. When it was loaded to 14 kN (0.17 P_u), the specimen had a small sudden drop in load due to the existence of bolt holes. After 20 kN (0.25 P_u), the specimen had a load bearing plateau, and the displacement increased by 2.2 mm. After 30 kN (0.37 P_u), the tensile stiffness of the specimen recovered and the load increased linearly. When the load was continued, the specimen continued to make a fiber tearing sound. When the peak load of the specimen was 80.8 kN (P_u), the specimen was suddenly damaged.

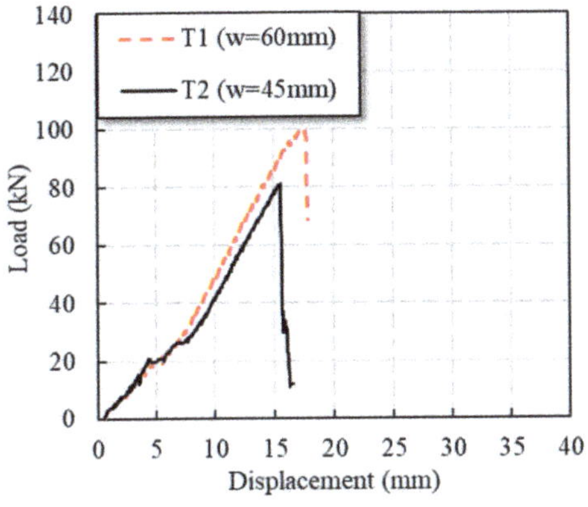

Figure 7. Load–displacement curves of specimens with different plate widths.

Table 1 shows that the ultimate bearing capacity of the specimens increased with the increase in w/φ, and the increase was more than 20%. With the increase in w/φ, the failure mode of the specimen changed: the GFRP plate of the T1 specimen has an obvious interlaminar shear effect when it is tensile, and its failure mode damage is that of bearing failure; the GFRP plate of the T2 specimen fractures along the bolt hole with interlaminar shear, and its failure mode is that of shear-out failure. It can be found from Figure 7 that the stiffness of the specimen with higher w/φ was improved to a certain extent in the later stage of loading. Compared with shear-out failure, the effect of bearing failure results in higher load capacity.

3.2.3. Edge Distance of Plate

The end distance is determined according to the size of the hole diameter. The control end distance/hole diameter (e/φ) ratio is 3.3 at minimum. The test phenomenon of the T3 and T4 specimens was similar to that of T1, but the T3 and T4 specimens had post-peak load responses until gradual failure, which was caused by the automatic stop of laboratory instruments in the T1 specimens. It can be seen from Table 1 and Figure 8 that the change in GFRP plate end distance between approximately 40 and 60 mm had little effect on load response, meaning that it can almost be ignored.

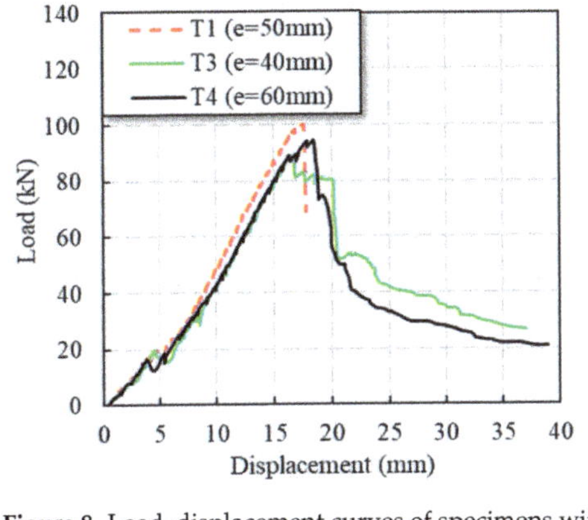

Figure 8. Load–displacement curves of specimens with different end distances.

3.2.4. Thickness of Plate

It can be seen from Table 1 that the thickness of the GFRP plate can improve the P_u of specimens, which can also be found in Figure 9. The failure mode of T6 is tensile failure. The peak load of specimen T1 was 62.3% higher than that of T6. Figure 9 shows that the stiffness of the T1 specimen was significantly increased compared with the T6 specimen after 40 kN. However, the T1 specimen did not show post-peak response, whereas the load of specimen T6 gradually decreased after the peak load.

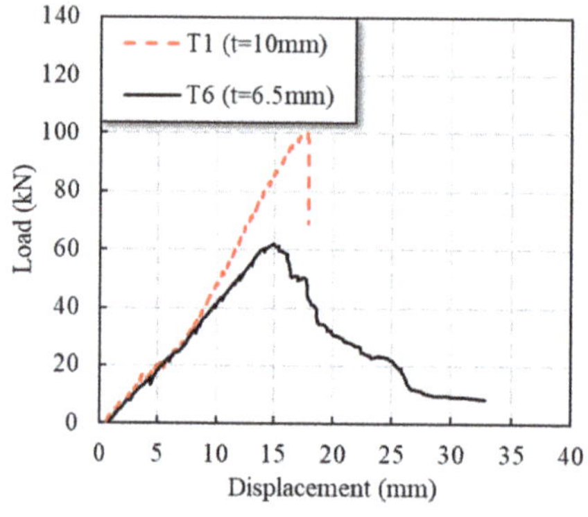

Figure 9. Load–displacement curves of specimens with different plate thicknesses.

3.2.5. Adhesive

As Figure 10a shows, during the loading process of the T5 specimen, before 10 kN (0.10 P_u), the load basically continued to rise linearly, and the initial tensile stiffness of the specimen was small; after 10 kN, the stiffness of the specimen increased slightly. When the load reached 49 kN (0.50 P_u), the adhesive layer was shear damaged, and the specimen made a loud cracking sound. At the same time, the load suddenly dropped to 33 kN (0.34 P_u), and then the stiffness of the specimen recovered, and the load continued to rise linearly. With continued loading, the specimen made a crisp sound of fiber tearing; when approaching the peak load of the specimen, the fiber emitted a continuous fiber tearing sound; after the peak load of 97.9 kN (P_u), the load suddenly dropped to 64 kN, and then the load decreased in a curve. In the later stage of load reduction, the specimen showed good ductility.

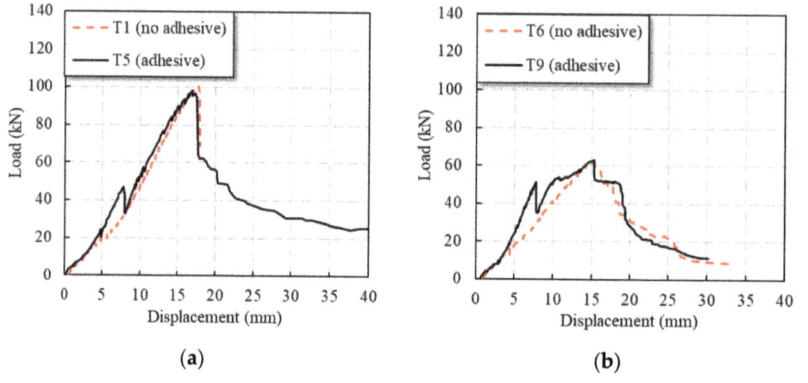

Figure 10. Load–displacement curves for comparison of the bolted and bolted–bonded hybrid connection: (**a**) thick plate, and (**b**) thin plate.

As Figure 10a shows, the stiffness of the T9 specimen was relatively small before 10 kN (0.16 P_u). After 10 kN, the stiffness of the specimen increased significantly and the load increased linearly. When the load was 50 kN (0.79 P_u), due to the shear failure of the adhesive layer, there was a loud cracking sound, and the load suddenly dropped to 35 kN. Then, the load continued to rise, but the stiffness of the specimen decreased. When the load was 54 kN (0.85 P_u), the specimen slipped, and the tensile stiffness of the specimen was further reduced. When the load was continued, the specimen emitted a continuous fiber tearing sound. After reaching the peak load of 63.2 kN (P_u), the load decreased, where two descending plateaus were observed.

It can be seen from Table 1 and Figure 10 that the interface adhesive did not increase the specimen's P_u. The initial stiffness of the specimen was significantly increased. When the load reached the shear strength of the adhesive layer, the load of the specimen decreased obviously. In the later stage of loading, the specimen mainly depended on the bolt to bear the load, so the P_u of the specimen was basically the same as the pure bolted connection. The specimens T6 and T9 have similar shear strength, but it can be clearly observed that the adhesive layer failure occurred when T9 was about 50 kN. At the same time, comparing the glued specimens in the two groups of thick and thin plates (the same plate width and the same adhesive layer area), it can be found that the adhesive layer shear failure occurred in the specimens at about 50 kN, indicating that the shear bearing capacity of the adhesive layer in the same area was the same.

3.3. *Numerical Simulation of Double-Lap Direct Shear Test*

3.3.1. Configuration of Finite Element Model of Double-Lap Direct Shear Test

A finite element (FE) model of the GFRP bolted joint was established in this study by using Abaqus, as shown in Figure 11. The bolt shank, nut, washer and GFRP plates were modeled according to the dimensions in Table 1 in the FE model, where a similar method has been shown in [48]. A reference point RP-1 was set on one end of the FE model of the GFRP bolted joint, and the reference point was coupled with the end surface, and the other end was constrained by consolidation.

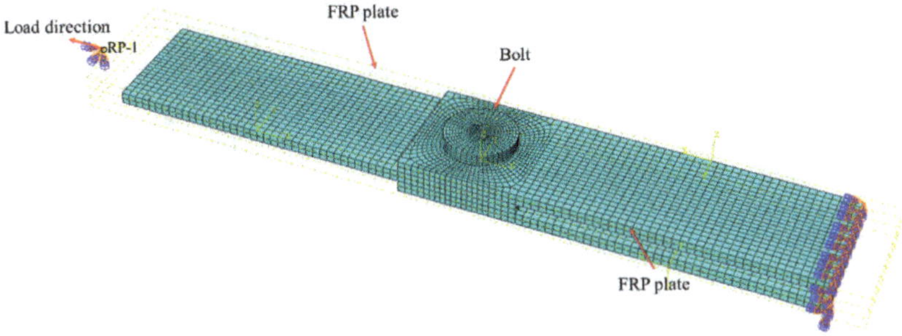

Figure 11. FE model of double-lap shear test of GFRP bolted joints.

The C3D8R element was used for the overall structure of the bolt, the density was 7850 kg/m^3, the Young's modulus was 200 GPa and the Poisson's ratio was 0.1. The SC8R element was used for the GFRP plate, the density was 2700 kg/m^3, the longitudinal Young's modulus was 46 GPa, the transversal Young's modulus was 9.0 GPa, and the in-plane shear modulus was 4.7 GPa, according to the manufacturer. In the established general contact, the friction coefficient of tangential behavior was selected as 0.1, and the normal behavior was selected as 'hard' contact and allowed to separate after contact. The mesh refinement was performed on the part where the bolt was in contact with the GFRP plate, increasing the accuracy of the mesh sizes to 1mm. In order to divide a suitable and easy-to-calculate mesh, the bolt was divided into a hexahedral structure as a whole. The GFRP plate was divided using the advanced algorithm in the hexahedral swept division and allowed the use of mesh mapping where appropriate.

The loading process was divided into two stages that were the same as the experimental test. First, we applied the preload and then applied monotonically increasing displacement at the loading end. The loading was solved by the dynamic explicit solver. The progressive failure of GFRP followed the Hashin damage initiation criteria [49]. Cohesive elements (zero-thickness elements) were employed to simulate the bonding layer in bolted–bonded specimens. In cohesive behavior, Knn was 30,000, Kss was 15,000, and Ktt was 15,000. In the damage attribute setting, the secondary tension criterion was selected. The damage of the cohesive element was output in the solution result.

3.3.2. Comparison between FE Model and Experimental Results of Double-Lap Direct Shear Test

The FE simulation results (Figure 12) show that the stress concentration phenomenon occurred in the 45° direction of the hole in the GFRP bolt connection. The failure mode matched well with the experimental results, as shown in Figure 13. In the experimental observations, the failure modes of T7 are both bearing failure and shear-out failure, which was also found in the results of numerical simulations, as shown in Figures 12 and 13. The simulated failure process can be extracted frame by frame, providing a more complete picture of the damage process of the specimen than can be achieved in a test. It should be noted that a deformation magnification factor of 8.5 times was selected in Figures 12 and 13.

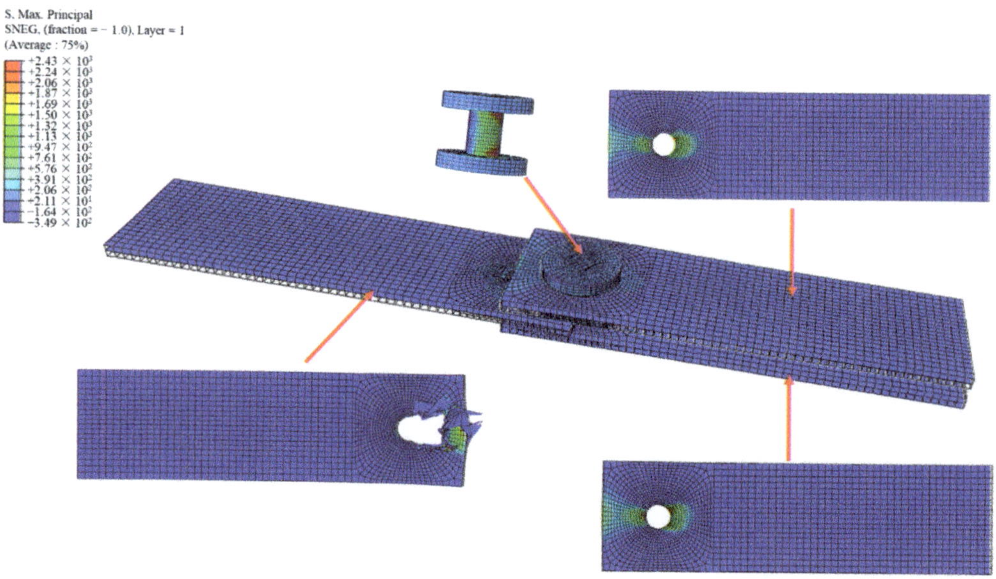

Figure 12. Maximum principal stress cloud of double−lap shear test of GFRP bolted joints.

Figure 14 shows that the results of the numerical simulation were close to the results of the experimental test in terms of the maximum load. However, the displacement corresponding to the maximum test load was about 5 mm larger than the simulation results, which might be due to the initial gap between the loading head and the GFRP plate specimen.

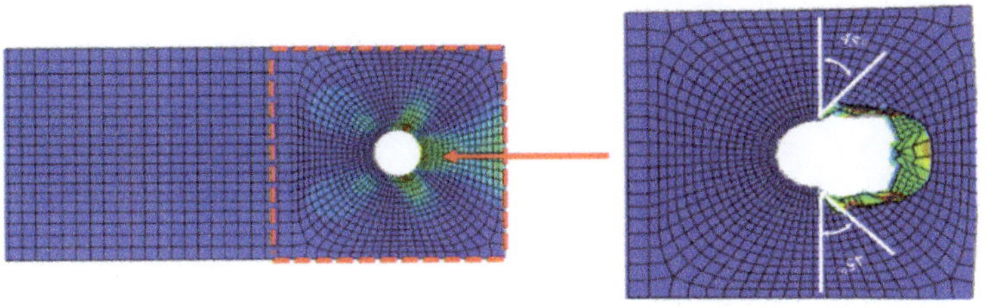

(**a**)

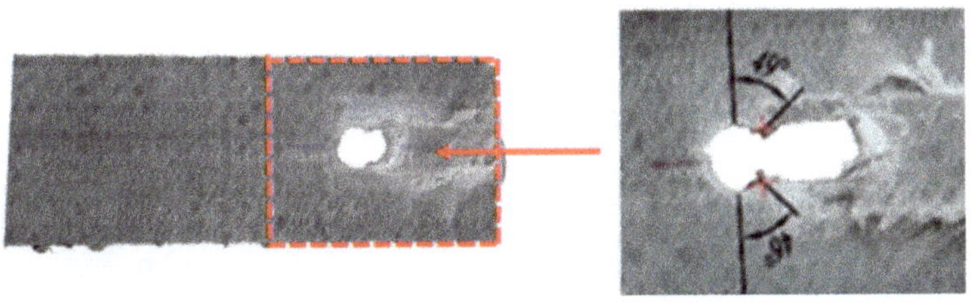

(**b**)

Figure 13. Comparison of (**a**) shear stress predicted by FE model, and (**b**) shear out failure mode of specimen T7.

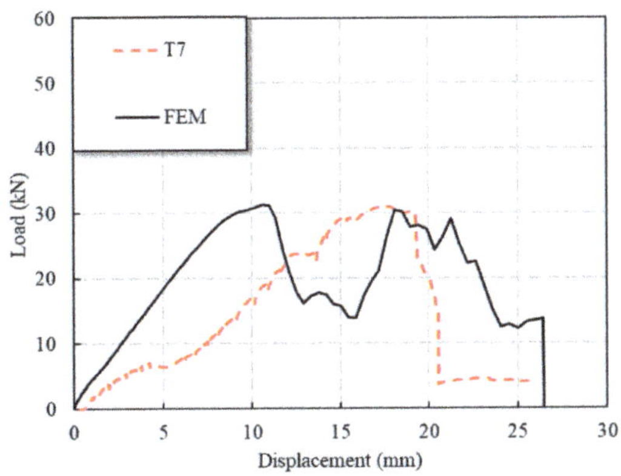

Figure 14. Comparison of FEM and test results of load response of specimen T7.

3.3.3. Analysis of Progressive Failure of Double-Lap Direct Shear Test Specimens

Referring to the numerical model of the T7 specimen, the numerical model of the double shear single bolt with adhesives was established. As can be seen from Figure 15, the FE model of specimens with adhesives showed a higher initial stiffness of the specimen, and the load was significantly reduced after the failure of the adhesive layer. The load response predicted by the FE model was in good agreement with the experimental observations of the double-shear connection specimens T5 and T9.

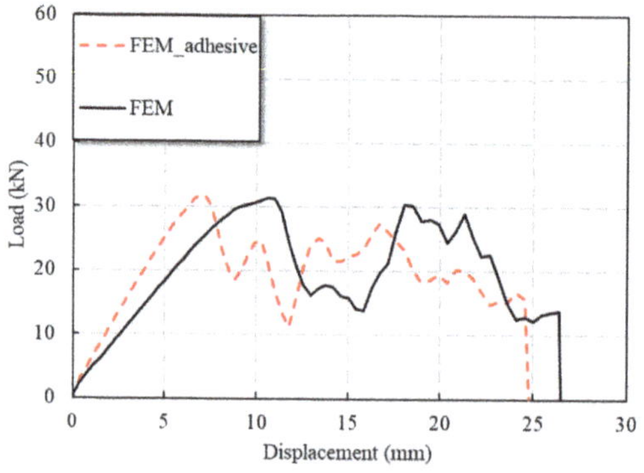

Figure 15. Comparison of FE model results: bolted and bolted–bonded hybrid connection.

Figure 16 demonstrates that the adhesive on the loading end side first failed (see Point A in the load–displacement curve in Figure 16), and then the failure of adhesive near the fixed end was observed. As the displacement increased, the interfacial damage on one side of the fixed end extended to both sides and the middle. With the increase in displacement, the load reached a plateau after the peak of the load. This plateau, according to the authors' analysis, is due to the progressive failure of the interfacial adhesive, as can be shown by Points F, G, and H in the load–displacement curve in Figure 16. This process was not apparent during the experiments, but some acoustic activities were heard during the test. Finally, the adhesive layer failed completely at the end of the post-peak plateau of the load (see Point I in the load–displacement curve in Figure 16).

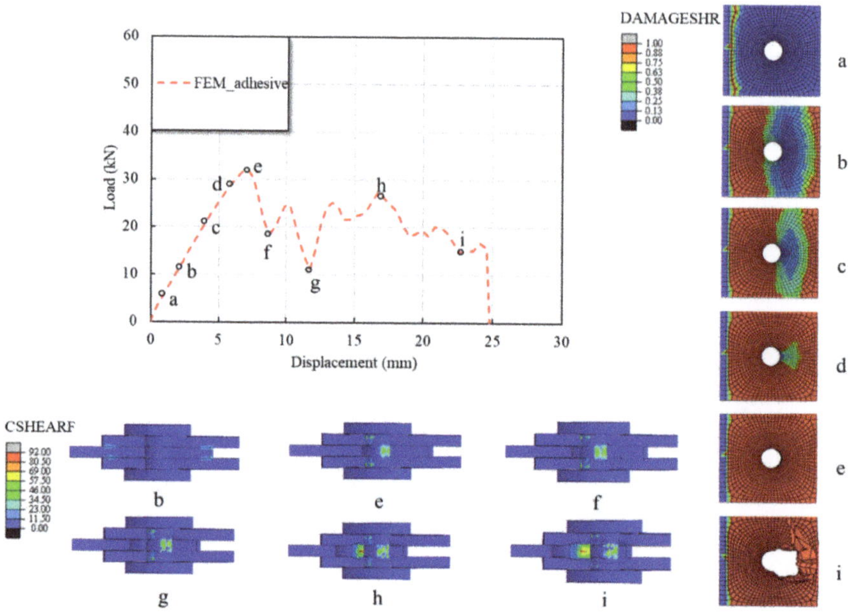

Figure 16. Progressive failure of the bolted–bonded hybrid connection specimen.

4. Results and Analysis of Full-Scale GFRP Joint Test

4.1. Tests Results and Observations of Full-Scale GFRP Joint Test

The load–midspan displacement curves of specimen BJ, GP and IGP are shown in Figure 17. For specimen BJ, the first crack formed at the web–flange junction of the horizontal chord when the load reached 33.8 kN. At 37.4 kN, longitudinal shear cracks appeared between the upper row of bolts in the rear grooved web chord joint, while shear cracks formed at the hinge of the left diagonal chord. At 41.8 kN, multiple parallel longitudinal web shear cracks formed between the bolts of the front grooved web chord. When the bottom chord web chord reached the ultimate bearing capacity, the load dropped suddenly, and the end of the vertical chord extended out of the core area.

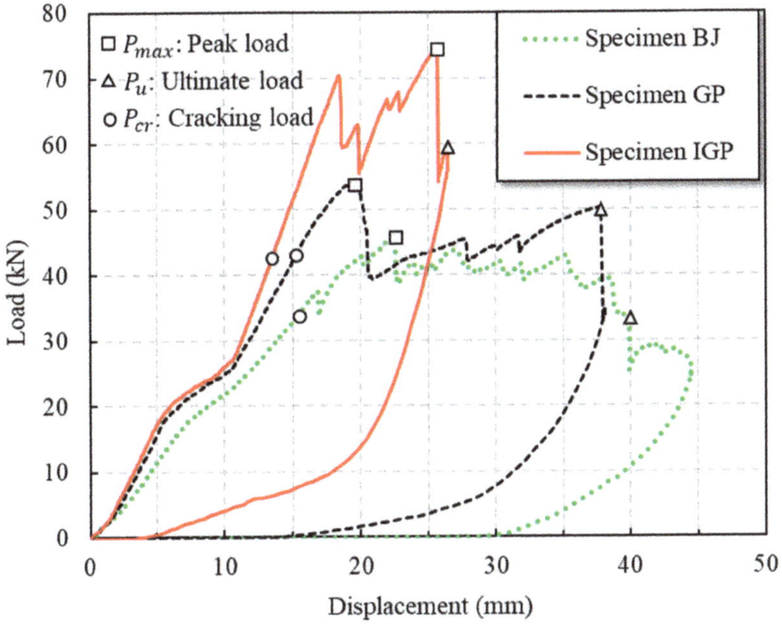

Figure 17. Load–displacement curve of GFRP truss joints.

For specimen GP, cracks formed at the midspan flange of the front grooved web chord when the load reached 44.2 kN. With the increase in load, the top flange of the crack member collapsed and the deformation of the flange increased gradually. When the load reached 53.6 kN, the connection between the left web and the bottom plate of the front grooved web chord was fractured. After that, with the continuous increase in load, the vertical load was mainly carried by the hinged part of the diagonal chord. When the load decreased to around 48 kN, the pull-out failure occurred at the hinge of the right diagonal chord, resulting in the overall failure of the joint.

Specimen IGP experienced its first crack at the load of 42.33 kN at the web–flange junction of the horizontal chord. With the increase in load, the crack propagated along the vertical direction and finally resulted in overall failure at the ultimate load of 54.12 kN. The peak load of specimen IGP (74.47 kN) is 62% higher than the bolted one (specimen BJ), and 38.8% higher than the typical gusset plate joint (specimen GP).

Comparing the load–displacement curves of specimen BJ, GP and IGP (Figure 17), it can be seen that the typical bolted gusset plate (specimen GP) did not significantly increase the maximum load carrying capacity of the members. This might be caused by the fact that the lower chord pultruded profile was divided into two halves along the middle of the specimen in which the novel bolted gusset plate was incorporated. However, the inclusion of the novel bolted gusset plate significantly reduced the cracking and increased the ductility characteristics of the specimens by 21.62%. The employment of the proposed gusset plate substantially improved the peak load but reduced the ductility when compared with the rest two specimens. The reason for this improvement, according to the authors' analysis, is that the proposed IGP works as a strengthening component of the joint and allows the bolts to work together to withstand the load. It is noted that ductility was estimated from the displacement to the point where the load dropped sharply after reaching the peak load.

The test results of specimen BJ, GP and IGP are shown in Table 3, where P_{cr} and δ_{cr} are the load and displacement at the initial cracking of the vertical chord, respectively. P_{d1} and δ_{d1} are the load and displacement when stiffness degradation occurred at the vertical chord, respectively. P_{d2} and δ_{d2} are the load and displacement when stiffness degradation occurred, respectively. P_{max} and δ_{max} are the peak load and corresponding displacement of the vertical chord. P_u and δ_u are the ultimate load and corresponding displacement, respectively. Failure mode (a) represents the longitudinal shear of the web; (b) represents the shear and emergence of the end of the diagonal chord; (c) represents the failure of the bottom chord; (d) represents the buckling of gusset plate; (e) represents the bending of the bolt under compression; and (f) represents lower chord buckling.

Table 3. Summary of static performance test results of GFRP pultruded truss joints.

Specimen	Failure Mode	P_{cr} /kN	δ_{cr} /mm	P_{d1} /kN	δ_{d1} /mm	P_{d2} /kN	δ_{d2} /mm	P_{max} /kN	δ_{max} /mm	P_u /kN	δ_u /mm
BJ	(a) (b) (d)	33.75	15.36	17.43	7.25	22.53	10.48	45.78	22.44	36.33	38.73
GP	(b) (c) (d)	44.22	15.53	17.95	5.62	25.57	10.53	53.66	18.97	33.48	37.97
IGP	(a) (b) (c)	42.23	13.40	17.25	4.99	27.11	10.63	74.47	25.68	54.12	25.76

4.2. Failure Modes of Full-Scale GFRP Joint Test

There were four typical failure modes in the static performance test of GFRP pultruded truss joints, as shown in Figure 18. Among them, specimen BJ showed shear out at the end of the diagonal chord, collapse of the top plate of the lower chord and buckling of the lower chord (see Figure 18e). Specimen GP was characterized by longitudinal shear of web, shear out of diagonal chord end and bottom chord buckling (see Figure 18f). Specimen IGP had a failure mode of shear failure at the web–flange junction (see Figure 18g).

After disassembling the failure joints, the distribution of the static failure cracks of the bottom chord web is shown in Figure 19. A large number of longitudinal shear cracks appeared in the bottom web chord under the action of vertical load, and some large cracks extending into the core area of the joint directly led to the failure of the web chord. There was no full-length crack through the bottom chord. It should be noted that the distribution of all shear cracks was biased towards one side of the member. This was because the occurrence of shear cracks significantly reduced the stiffness of one side of the member, resulting in a small rotation of the joint around the core area, resulting in asymmetric failure on both sides of the lower chord web chord. The load orthogonal to the pultrusion direction of GFRP controls the mechanical resistance of the bolted connection, so it is recommended to add an adhesive between gusset plates and GFRP truss members to improve the capacity.

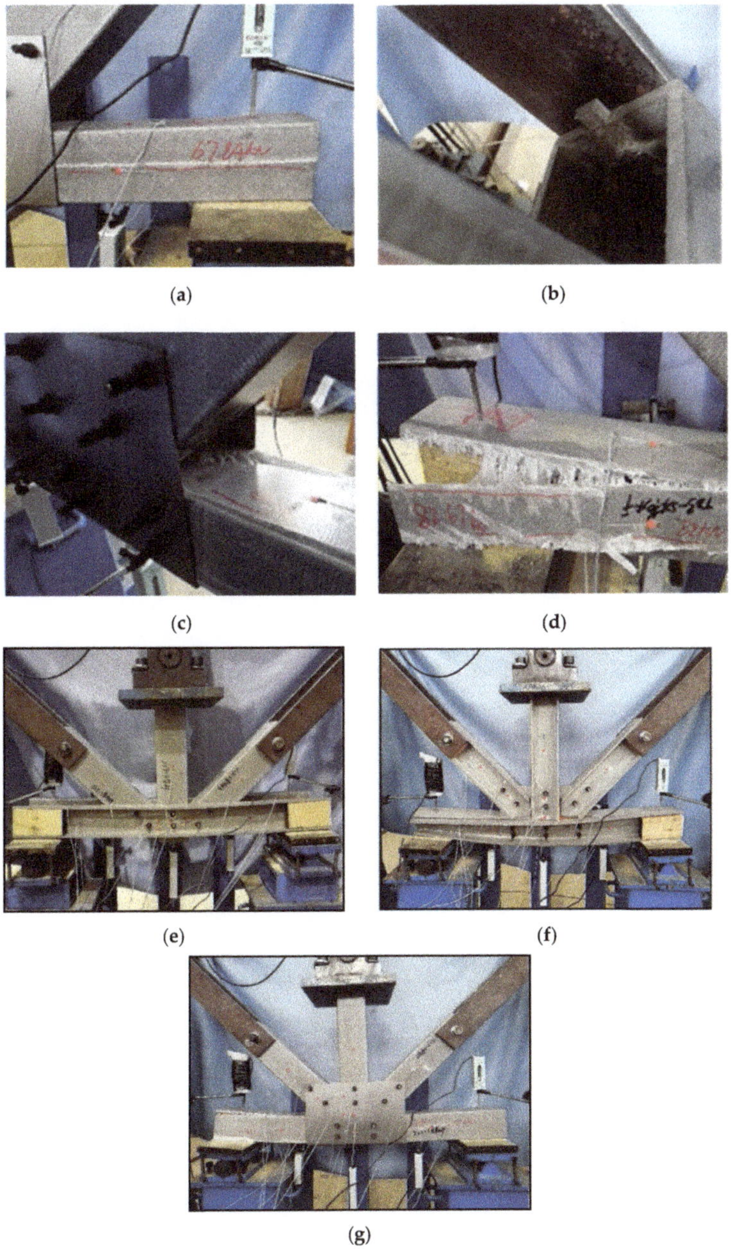

Figure 18. Failure modes of GFRP truss joints: (a) shear-out of webs, (b) shear-out of diagonal chords, (c) crush of top flanges, (d) separation of top flange and webs, (e) failed specimen BJ, (f) failed specimen GP, and (g) failed specimen IGP.

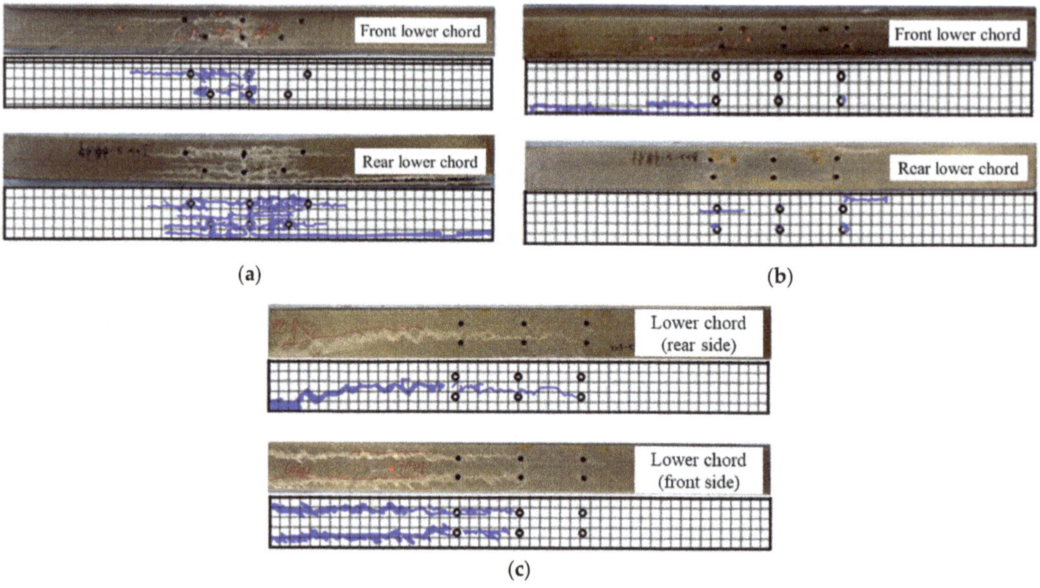

Figure 19. Distribution diagram of cracks in the lower chord: (**a**) specimen BJ, and (**b**) specimen GP, and (**c**) specimen IGP.

5. Comparison between Double-Lap Direct Shear Test and Full-Scale GFRP Joint Test

Both the double-lap direct shear test and the joint test can reflect the failure characteristics of GFRP pultruded bolted joints to a certain extent. Through comparison, it can be found that in the double-lap direct shear test and joint test, the failure of bolted joints mostly involves shear failure, and their cracks initiate at 45° around the hole and expand along the fiber pultrusion direction. However, there are more failure modes in the full-scale GFRP pultruded profile truss joint test, such as the collapse of the bottom chord roof and the failure of the web–flange connection, so it is necessary to carry out the full-scale test.

In terms of the load transferring mechanism (Figure 20), the double-lap direct shear test specimen can represent the loading transferring from vertical and diagonal chords to the gusset plates and horizontal truss members. However, most failure modes occurred at the webs of the horizontal GFRP member, which was not well represented by the direct shear test. So, it can be concluded that the direct shear test can better replace the connecting capacity of vertical and diagonal chords.

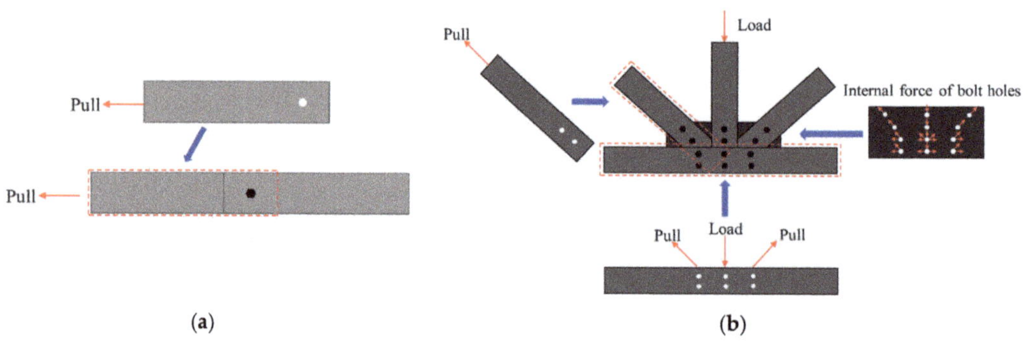

Figure 20. Failure modes of GFRP truss joints: (**a**) double-lap shear test, and (**b**) full-scale GFRP joint test.

6. Conclusions

In this study, a novel integrated gusset plate connection was proposed for GFRP truss structures. The static tests of nine GFRP bolted connections and three full-scale GFRP pultruded profile truss joints were carried out. The effects of the number of bolts, the geometric parameters of the plate, and the interface adhesive on the bearing capacity and failure mode of GFRP bolted connections were studied. Through experiments and theoretical analysis, the following conclusions can be drawn:

1. The peak load of the specimen with the proposed integrated gusset plate is 62% higher than the bolted one, and 38.8% higher than the typical gusset plate joint. As a comparison, the typical gusset plate can delay the formation of initial cracks and substantially reduce cracks, but it does not significantly improve the bearing capacity of joints.
2. The bolted–bonded hybrid connection has 50% higher shear stiffness and 27% higher ductility compared with bolted joints, so the bolted–bonded hybrid connection is recommended to be employed together with the proposed integrated gusset plate.
3. Bearing failure is the primary failure mode for most specimens. A crack formed at the 45° direction around the hole and propagated along the pultrusion direction until failure. It is recommended that more fibers shall be added to the transversal direction of the FRP members to be connected.
4. Compared with bolted GFRP connections, the failure of GFRP pultruded profile truss joints is similar to the phenomenon in practical engineering of GFRP truss structures. The double-lap direct shear test specimen can represent the loading transferring from vertical and diagonal chords to the gusset plates and horizontal truss members.
5. This research showed the advantages of using integrated gusset plates of the joints of the GFRP truss. It should be noted that GFRP can be used in field applications to replace the steel gusset plate used in this experiment considering durability requirements.

Author Contributions: Methodology, Y.C., X.Z., M.D. and Y.F.; testing and validation, Y.F.; simulation, Y.C.; writing—original draft preparation, X.Z. and Y.C.; methodology, writing—reviewing and editing, X.Z.; resources and supervision, X.Z. and G.L.; All authors have read and agreed to the published version of the manuscript.

Funding: This study was funded by the National Natural Science Foundation of Jiangsu Province, China (grant number BK20210620) and Natural Science Foundation of the Higher Education Institutions of Jiangsu Province, China (grant number 21KJB560007).

Institutional Review Board Statement: Not applicable.

Informed Consent Statement: Not applicable.

Data Availability Statement: Data sharing not applicable.

Acknowledgments: Many thanks go to the Engineering Training Center of Nanjing Forestry University and the Key Laboratory of Concrete and Prestressed Concrete Structures of the Ministry of Education of Southeast University for their help in the lab test.

Conflicts of Interest: The authors declare no conflict of interest.

Nomenclature

e	end distance of bolted pultruded GFRP specimen;
FRP	fiber-reinforced polymer;
GFRP	glass fiber-reinforced polymer;
IGP	integrated gusset plate;
l	length of pultruded GFRP specimen;
p	longitudinal spacing of bolts;
t	thickness of pultruded GFRP specimen;
w	width of pultruded GFRP specimen;
φ	hole diameter;
P_{bolt}	ultimate load of a single bolt;

P_u	load at the ultimate point;
P_{cr}	load at the initial cracking point;
P_{d1}	load at the stiffness degradation starting point;
P_{d2}	load at the stiffness degradation end point;
P_{max}	load at the peak point;
δ_{cr}	displacement at the initial cracking point;
δ_{d1}	displacement at the stiffness degradation starting point;
δ_{d2}	displacement at the stiffness degradation end point;
δ_{max}	displacement at the peak point;
δ_u	displacement at the ultimate point.

References

1. Camanho, P.; Matthews, F. Stress analysis and strength prediction of mechanically fastened joints in FRP: A review. *Compos. Part A Appl. Sci. Manuf.* **1997**, *28*, 529–547. [CrossRef]
2. Feo, L.; Marra, G.; Mosallam, A. Stress analysis of multi-bolted joints for FRP pultruded composite structures. *Compos. Struct.* **2012**, *94*, 3769–3780. [CrossRef]
3. Li, C.; Guo, R.; Xian, G.; Li, H. Innovative compound-type anchorage system for a large-diameter pultruded carbon/glass hybrid rod for bridge cable. *Mater. Struct.* **2020**, *53*, 73. [CrossRef]
4. Li, C.; Yin, X.; Liu, Y.; Guo, R.; Xian, G. Long-term service evaluation of a pultruded carbon/glass hybrid rod exposed to elevated temperature, hydraulic pressure and fatigue load coupling. *Int. J. Fatigue* **2020**, *134*, 105480. [CrossRef]
5. Wang, Z.; Zhao, X.; Xian, G.; Wu, G.; Singh Raman, R.; Al-Saadi, S. Durability study on interlaminar shear behaviour of basal glass-and carbon-fibre reinforced polymer (B/G/CFRP) bars in seawater sea sand concrete environment. *Constr. Build. M* **2017**, *156*, 985–1004. [CrossRef]
6. Ferdous, W.; Manalo, A.; Salih, C.; Yu, P.; Abousnina, R.; Heyer, T.; Schubel, P. Behaviour of polymer filled composites for n... polymer railway sleepers. *Polymers* **2021**, *13*, 1324. [CrossRef]
7. Al-Fakher, U.; Manalo, A.; Ferdous, W.; Alajarmeh, O.; Aravinthan, T.; Zhuge, Y.; Bai, Y.; Edoo, A. Shear behaviour of hollow precast concrete-composite structures. *Mater. Struct.* **2021**, *54*, 84. [CrossRef]
8. Feng, P.; Wang, J.; Wang, Y.; Loughery, D.; Niu, D. Effects of corrosive environments on properties of pultruded GFRP plates. *Compos. Part B Eng.* **2014**, *67*, 427–433. [CrossRef]
9. Feng, P.; Li, Z.; Wang, J.; Liu, T. Novel joint for pultruded FRP beams and concrete-filled FRP columns: Conceptual and experimental investigations. *Compos. Struct.* **2022**, *287*, 115339. [CrossRef]
10. Hizam, R.; Manalo, A.; Karunasena, W.; Bai, Y. Behaviour of pultruded GFRP truss system connected using through-bolt with mechanical insert. *Compos. Part B Eng.* **2019**, *168*, 44–57. [CrossRef]
11. Imperiale, V.; Cosentino, E.; Weaver, P.; Bond, I. Compound joint: A novel design principle to improve strain allowables of FRP composite stringer run-outs. *Compos. Part A Appl. Sci. Manuf.* **2010**, *41*, 521–531. [CrossRef]
12. Zou, X.; Wang, J. Experimental study on joints and flexural behavior of FRP truss-UHPC hybrid bridge. *Compos. Struct.* **2018**, *203*, 414–424. [CrossRef]
13. Fang, H.; Xu, C.; Sun, H.; Liu, W. Experiment on bearing capacity of bolted joints for pultruded composite square profile. *Acta Mater. Compos. Sin.* **2014**, *31*, 809–817. (In Chinese)
14. Ma, Y.; Zhao, Q. Analysis of the bonded bolted hybrid composite joints' carrying capacity. *Acta Mater. Compos. Sin.* **2011**, *28*, 225–230. (In Chinese)
15. Ju, S.; Jiang, D.; Shenoi, R.; Xiao, J. Flexural properties of lightweight FRP composite truss structures. *J. Compos. Mater.* **2011**, *45*, 1921–1930. [CrossRef]
16. McCarthy, C.; McCarthy, M.; Lawlor, V. Progressive damage analysis of multi-bolt composite joints with variable bolt-hole clearances. *Compos. Part B Eng.* **2005**, *36*, 290–305. [CrossRef]
17. Keller, T.; Bai, Y.; Vallée, T. Long-term performance of a glass fiber-reinforced polymer truss bridge. *J. Compos. Constr.* **2007**, *11*, 99–108. [CrossRef]
18. Keller, T.; Schaumann, E.; Vallée, T. Flexural behavior of a hybrid FRP and lightweight concrete sandwich bridge deck. *Compos. Part A Appl. Sci. Manuf.* **2007**, *38*, 879–889. [CrossRef]
19. Russo, S. First investigation on mixed cracks and failure modes in multi-bolted FRP plates. *Compos. Struct.* **2016**, *154*, 17–30. [CrossRef]
20. Lu, Z.; Yang, Q.; Lin, J. Study on influence of adhesive layer size on performance for single-lap bonding joint. *China Adhes.* **2009**, *18*, 1–3. (In Chinese)
21. Yang, C.; Huang, H.; Tomblin, J.; Sun, W. Elastic-plastic model of adhesive-bonded single-lap composite joints. *J. Compos. Mater.* **2004**, *38*, 293–309. [CrossRef]
22. Khashaba, U.; Sallam, H.; Al-Shorbagy, A.; Seif, M. Effect of washer size and tightening torque on the performance of bolted joints in composite structures. *Compos. Struct.* **2006**, *73*, 310–317. [CrossRef]
23. Chishti, M.; Wang, C.; Thomson, R.; Orifici, A. Numerical analysis of damage progression and strength of countersunk composite joints. *Compos. Struct.* **2012**, *94*, 643–653. [CrossRef]

24. Russo, S. On failure modes and design of multi-bolted FRP plate in structural joints. *Compos. Struct.* **2019**, *218*, 27–38. [CrossRef]
25. Sun, H.; Chang, F.; Qing, X. The response of composite joints with bolt-clamping loads, part I: Model development. *J. Compos. Mater.* **2002**, *36*, 47–67. [CrossRef]
26. Sun, H.; Chang, F.; Qing, X. The response of composite joints with bolt-clamping loads, part II: Model verification. *J. Compos. Mater.* **2002**, *36*, 69–92. [CrossRef]
27. Wu, Z.; Islam, S.; Said, H. A three-parameter bond strength model for FRP-concrete interface. *J. Reinf. Plast. Compos.* **2009**, *28*, 2309–2323. [CrossRef]
28. Yang, K.; Bai, Y.; Ding, C.; Kong, S. Comparative study on mechanical performance of bolted joints with steel and fibre reinforced polymer bolts. *J. Build. Eng.* **2021**, *41*, 102457. [CrossRef]
29. Gu, Y. Strength Analysis and Influencing Factors of Mechanical Joints of Composite Materials. Ph.D. Thesis, Northwestern Polytechnical University, Xi'an, China, 2017. (In Chinese)
30. Wu, C.; Zhang, Z.; Xu, N.; Tan, L. Experimental study on the group effect of GFRP-timber bolted connection in tension. *China Civ. Eng. J.* **2021**, *54*, 38–48. (In Chinese)
31. Wu, C.; Bai, Y.; Mottram, J. Effect of elevated temperatures on the mechanical performance of pultruded FRP joints with a single ordinary or blind bolt. *J. Compos. Constr.* **2016**, *20*, 04015045. [CrossRef]
32. Abdelkerim, D.; Wang, X.; Ibrahim, H.; Wu, Z. Effect of Connection Techniques on the Static and Fatigue Performance of Pultruded Basalt FRP Multibolted Joints. *J. Compos. Constr.* **2020**, *24*, 04020046. [CrossRef]
33. Okutan, B.; Karakuzu, R. The strength of pinned joints in laminated composites. *Compos. Sci. Technol.* **2003**, *63*, 893–905. [CrossRef]
34. Chen, S. Experimental study on mechanical properties of GFRP pedestrian truss bridge. Ph.D. Thesis, Southeast University, Nanjing, China, 2017. (In Chinese)
35. Ascione, F.; Lamberti, M.; Razaqpur, A.; Spadea, S.; Malagic, M. Pseudo-ductile failure of adhesively joined GFRP beam-column connections: An experimental and numerical investigation. *Compos. Struct.* **2018**, *200*, 864–873. [CrossRef]
36. Russo, S. Damage assessment of GFRP pultruded structural elements. *Compos. Struct.* **2013**, *96*, 661–669. [CrossRef]
37. Russo, S. Shear and local effects in All-FRP bolted built-up columns. *Adv. Struct. Eng.* **2015**, *18*, 1227–1240. [CrossRef]
38. Castellaro, S.; Russo, S. Dynamic characterization of an all-FRP pultruded construction. *Compos. Struct.* **2019**, *218*, 1–14. [CrossRef]
39. Zhang, H.; Zhang, L.; Liu, Z.; Qi, S.; Zhu, Y.; Zhu, P. Numerical analysis of hybrid (bonded/bolted) FRP composite joints: A review. *Compos. Struct.* **2021**, *262*, 113606. [CrossRef]
40. Mutsuyoshi, H.; Nguyen, H.; Zatar, W.; Ishihama, T. Flexural behavior of pultruded hybrid fiber-reinforced polymer I-beams with bonded-and-bolted splice joints. *Transp. Res. Rec.* **2016**, *2592*, 45–55. [CrossRef]
41. Wu, C.; Feng, P.; Bai, Y. Comparative study on static and fatigue performances of pultruded GFRP joints using ordinary and blind bolts. *J. Compos. Constr.* **2015**, *19*, 04014065. [CrossRef]
42. Coelho, A.; Mottram, J. A review of the behaviour and analysis of bolted connections and joints in pultruded fiber reinforced polymers. *Mater. Des.* **2015**, *74*, 86–107. [CrossRef]
43. Turvey, G.; Cerutti, X. Flexural behaviour of pultruded glass fiber reinforced polymer composite beams with bolted splice joints. *Compos. Struct.* **2015**, *119*, 543–550. [CrossRef]
44. Hai, N.; Mutsuyoshi, H. Structural behavior of double-lap joints of steel splice plates bolted/bonded to pultruded hybrid CFRP/GFRP laminates. *Constr. Build. Mater.* **2012**, *30*, 347–359. [CrossRef]
45. Liu, T.; Feng, P.; Lu, X.; Yang, J.; Wu, Y. Flexural behavior of novel hybrid multicell GFRP-concrete beam. *Compos. Struct.* **2020**, *250*, 112606. [CrossRef]
46. Liu, T.; Liu, X.; Feng, P. A comprehensive review on mechanical properties of pultruded FRP composites subjected to long-term environmental effects. *Compos. Part B Eng.* **2020**, *191*, 107958. [CrossRef]
47. Standard of China Engineering Construction Standardization Association. *Technical Specification for Structure of Pultruded Fiber Reinforced Polymer Composites*; T/CECS 692-2020; China Construction Industry Press: Beijing, China, 2020.
48. Zhu, Y.; Zhu, H.; Gribniak, V. Analyzing the Sample Geometry Effect on Mechanical Performance of Drilled GFRP Connections. *Materials* **2022**, *15*, 2901. [CrossRef] [PubMed]
49. Hashin, Z. Failure criteria for unidirectional fiber composites. *J. Appl. Mech.* **1980**, *47*, 329–334. [CrossRef]

polymers

MDPI

Article

Study on Bending Creep Performance of GFRP-Reinforced PVC-Based Wood-Plastic Composite Panels

Bangbang Dai, Ruili Huo *, Kun Wang, Zhengqing Ma and Hai Fang

College of Civil Engineering, Nanjing Tech University, Nanjing 211816, China
* Correspondence: ruilihuo@njtech.edu.cn

Abstract: Wood-plastic composites (WPCs) are environment-friendly materials, which have broad application prospects in structures. They cannot be used for bearing structures because of poor mechanical performance and creep deformation. In order to enhance the mechanical behavior and decrease the long-term creep deformation, glass fiber reinforced plastics (GFRP) sheets and rebar reinforcement design methods are proposed. The bending static tests and creep performance tests of WPCs were conducted. The results showed that GFRP sheets and rebars improved the ultimate flexural loading capacity and deformation capacity by 257% and 165%, respectively, decreased the creep deflection effectively, and avoided shear failure. When the load level was very low, the creep deformation of WPC panels unreinforced, or reinforcement developed stably with time, and the damage did not occur within 1100 h. When the load increased to 80% of the ultimate load level, all specimens were damaged in the compression zone, the creep deformation increased quickly and unstably, bending shear failure of the unreinforced specimen occurred after 7 h, shear failure of the GFRP-sheets-reinforced specimen occurred after 1100 h, and the rebar-reinforced specimen failed after 720 h with excessive deflection deformation in the span. The reinforced effect of GFRP sheets is better. The creep strain growth rate of all specimens increased quickly at the first stage and gradually decreased at the second stage and tended to be stable. The creep calculation model was built based on the four-element model, which is simple and efficient and can make scientific and reasonable predictions of the two phases of structural transient creep and deceleration creep.

Keywords: GFRP reinforced; wood-plastic composites; bending creep; load level; four-element model

1. Introduction

Composite material structures are more popular in engineering applications due to their excellent physical and mechanical properties, especially under harsh environmental conditions [1–3]. PVC-based wood-plastic composites are a new type of environmentally friendly composite material made from wood fibers or plant fibers as reinforcing materials or fillers and polyvinyl chloride thermoplastics as the matrix, which are melt-mixed and then subjected to a molding and processing process [4]. They have the advantages of good processing performance, corrosion resistance and mechanical performance, wide resource of raw materials, energy saving and environmental protection, weather resistance, creep performance is higher than natural wood [5], and they have been widely used as structural materials in the field of garden architecture, exterior wall panels, outdoor decking, indoor and outdoor decoration, automotive interior trim, etc. However, WPCs are deficient in weather resistance, flame retardancy, mechanical strength, etc., especially their poor creep resistance, which is prominent in long-term use, affecting their application in the engineering field. GFRP has the advantages of high tensile strength, high modulus of elasticity, corrosion resistance, abrasion resistance, and low cost, which is also used to reinforce resin synthetic composites [6–8].

In recent years, researchers have conducted studies on the component materials' behavior [9–12], manufacture processes [13,14], structural forms [15–17], and mechanical performance [18–21] of WPCs. As the resin material exhibits viscoelastic properties, the resin molecules are displaced along the direction of the stress field under continuous loading, and the molecular chains are gradually straightened from their original bent state to result in creep [22,23]. GFRP creep is mainly divided into the following processes: ① local elongation of glass fibers; the resin has a restraining effect on the fiber; ② local fracture of glass fibers; the resin around the fiber transfers stress, and the resin creep in the high-stress area increases; ③ when the shear stress at the fiber resin interface is parallel to the fiber filaments, the interface damage occurs in single fiber bundles under the effect of continuous stress, and combined with the effect of their own defects, leads to the development of cracks; ④ a number of fiber bundle fractures; the creep rate keeps increasing [24–26].

Results have been obtained on the creep properties of GFRP under long-term loading. Dutta et al. [27] studied the tensile and compressive properties of GFRP rebars under continuous loading and high temperature, analyzed the effect of the superposition effect of temperature and time on the creep properties of GFRP rebars, and proposed a simple empirical model to predict the failure time of GFRP based on the equation fitted to the creep curve. Tannous et al. [28] found that the creep resistance is better and the deformation is smaller when the GFRP rebars are at consistently lower stress levels. Uomoto et al. [29] found that the creep deformation of GFRP rebars is large when the sustained stress level reaches 80% of the ultimate strength of the GFRP rebars. Guowei Li et al. [30,31] found that GFRP creep occurs only when the load level reaches 77% of the ultimate load, but creep deformation of GFRP was not found in the experiment, indicating that GFRP has excellent creep resistance under the effect of a long-term low load level, which is consistent with the results of Qi Sun [32]. Yihua Cui et al. [33] concluded that the interaction of glass fiber, wood powder, and plastic formed a stable three-dimensional spatial structure, which existed in the plastic matrix as a strong skeleton and prevented the generation and expansion of cracks, thus improving the mechanical strength and impact resistance of the composites; the mechanical properties of WPCs show a trend of increasing and then decreasing with an increase in glass fiber content. Dong Guo et al. [34] found that the impact strength, flexural strength, and flexural modulus of the composites reached the maximum when the content of glass fiber was 30%, which were 121%, 174%, and 79% higher than those of the composites without glass fiber, respectively; the mechanical properties decreased when the glass fiber content increased to 40%. Jeamtrakull et al. [35] showed that the wear resistance of the composites was significantly improved by the addition of 10% glass fibers (length 3–13 mm), indicating that the addition of short-cut glass fibers could improve the wear resistance of the composites. Huang and Kim [36,37] prepared WPCs with a core/shell structure by co-extrusion, using short-cut glass fiber-reinforced high-density polyethylene for the shell and high-density polyethylene WPCs for the core. The results showed that the fibers in the shell reduced the linear expansion of the material, and the shell thickness and fiber content had different reinforcing effects on the core WPCs. Overall, the core/shell structure reduces the amount of fiber and achieves reinforcement at the same time. Zolfaghari et al. [38] developed WPCs reinforced with continuous untwisted glass fiber bundles and showed that the continuous glass fiber-reinforced WPCs have a better reinforcement effect than short-cut fibers, which is significant to solve the brittleness problem and expand the applications of WPCs. Pulngern et al. [39,40] investigated the mechanical performance of WPCs under different temperatures and loads; Sain et al. [41] improved the existing creep model and Findley index model for WPCs; Li et al. [42] investigated the flexural creep performance of glass fiber-reinforced polymer composite sandwich beams and proposed a life prediction method to predict the creep life of beams under arbitrary load levels. Xianling Tian et al. [43] studied the creep performance of WPCs under different loading ways; Yan Cao et al. [44] investigated the creep performance and creep model of molded poplar wood fiber (PWF)/HDPE composites; Huhu Du et al. [45] and Weiren Xie [46] simulated the creep behavior of WPCs using the multi-element model and power law model based on experimental results. Some technical standards were made through experimental studies on the creep performance of WPCs [47]. Some scholars have studied the effects of material factors, process factors, environmental factors, and service factors on the creep performance of WPCs [48–52]. Conventional composite components are highly susceptible to peeling damage at the interface between the face and core during fabrication and in service, which severely limits their lightweight and high-strength properties [53–55]. The creep deformation caused by long-term loading cannot be ignored when WPCs are used as structural members under constant environmental and load effects, and the creep deformation will lead to the reduction of structural bearing capacity or instability damage, and the damage caused by creep is one of the main forms of damage to the members, which directly affects the reliability and safety of the structure, so it is mainly used in the field of non-structural materials [56]. Most materials' creep will occur under long-term loading [57–59], which causes material, structural strength, and stiffness degradation [60], affecting the safety, serviceability, and durability of structures. GFRP has the characteristics of a light weight and high strength, corrosion resistance, fatigue resistance, and excellent creep resistance [61–63]. This paper used the GFRP reinforcement method to overcome the shortcomings of PVC-based WPCs, such as insufficient mechanical properties and large creep deformation, and carried out three-point bending creep performance tests under different load levels to provide theoretical references for the design of WPCs' loading structural members and improve the durability performance of structures.

2. Experiments

2.1. Raw Materials

Based on the application of WPCs, we chose the widely used PVC-based WPC materials to make specimens. The density was 0.82 g/cm^3, Young's modulus was 960 MPa, bulk modulus of elasticity was 533 MPa, yield strength was 9.7 MPa, and Poisson's ratio was 0.2, which were produced by Keju New Material Technology Co, Ltd. Anhui, China. The main components were poplar wood powder, PVC resin, foaming agent, coupling agent, etc., prepared by the co-extrusion molding process. The tensile strength of the GFRP sheet was 300 MPa, the modulus of elasticity was 20 GPa, Poisson's ratio was 0.15; the tensile strength of the GFRP rebar was 490 MPa and the modulus of elasticity was 42 GPa.

2.2. Samples' Design and Preparation

There are three types of specimens: WPCs, GFRP-sheets-reinforced WPCs, and GFRP-rebar-reinforced WPCs. The dimensions of specimens were: 630 mm × 80 mm × 36 mm, as shown in Figure 1. GFRP-sheets-reinforced WPCs were made of a double-layer PVC-based wood-plastic panel and GFRP sheet; the thickness of the GFRP sheet was 1.2 mm, which consists of an unsaturated polyester resin matrix and biaxially orthogonal (0, 90°) woven glass fiber cloth in one piece produced by vacuum introduction technology [64] and two layers of 0.6 mm GFRP sheets glued together. The PVC-based wood-plastic panels were cut according to the designed size and a margin of 10 mm was left on the length and width to facilitate the secondary processing of the specimen after the gluing was completed. Considering that the PVC-based WPC panels' surface was too smooth and affected the gluing quality during production, the polishing process was necessary. The specimens were cut after 48 h of curing. The GFRP-rebar-reinforced PVC-based wood-plastic panels were prepared by an embedding process, and the diameter of the GFRP threaded rebar was 8 mm. Firstly, longitudinal penetration grooves were engraved on the bottom surfaces of WPCs, then epoxy resin glue was filled in the groove, and the GFRP rebars were embedded in the opened groove, as shown in Figure 2.

2.3. Bending Static Load Test

Firstly, the three-point bending static load test of the reinforced and unreinforced specimens was carried out to determine their ultimate load capacity and in order to determine the load level of the bending creep performance test. The span sizes of the specimens were 576 mm, an MTS electronic universal testing machine was used in the test, and the loading rate was 2 mm/min by the method of controlled displacement. In order to prevent the local damage of the specimen surface during the loading process, a rubber mat of Shore A hardness 60 and size of 70 mm × 25 mm × 3 mm was placed at the span loading point on the upper surface of the specimen. The load-mid-span displacement curves were collected and recorded simultaneously during the loading process.

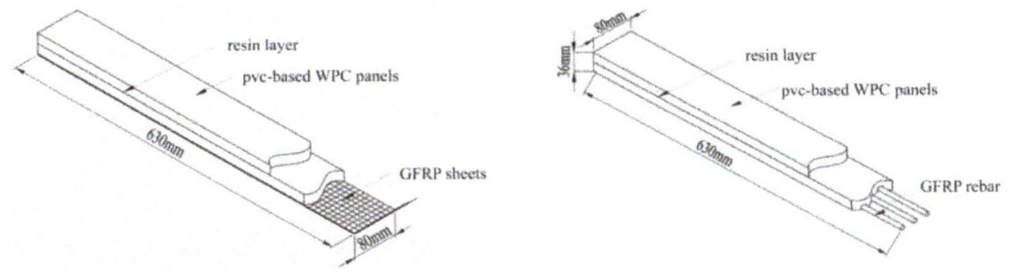

(**a**) Specimen reinforced with GFRP sheets (**b**) Specimen reinforced with GFRP rebar

Figure 1. Design of GFRP-sheets- and rebar-reinforced WPCs.

(**a**) WPC panel unreinforced (**b**)WPC panel reinforced with GFRP sheets (**c**) WPC panel reinforced with GFRP rebar

Figure 2. PVC-based WPC specimens.

2.4. Bending Creep Test

The bending creep test was carried out under constant temperature and humidity with three-point loading; the test device is as shown in Figure 3. The temperature and humidity controller could provide a stable temperature and humidity environment for the creep test; the temperature was kept at 23 °C ± 2 °C, the relative humidity was 50% ± 10%, and the displacement of the specimen was measured by dial indicator with the accuracy of 0.01 mm.

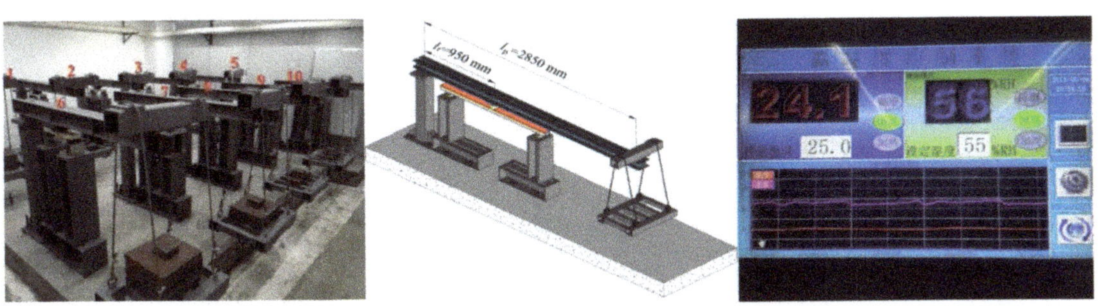

(**a**) Loading devices (**b**) Diagram of loading (**c**) Temperature and humidity controller

Figure 3. Three-point bending creep test loading devices.

Six load levels were set as the creep test design load: 30%, 40%, 50%, 60%, 70%, and 80% of the ultimate bearing capacity of the static test. The rubber mat was placed on the upper surface of the test piece span, adjusting the position of the dial indicator, so that its pointer was perpendicular to the lower surface of the specimen. When the loading end slowly reached the position of the rubber mat, the initial reading of the dial indicator was recorded. After that, the data were collected every 24 h. The test time for each specimen was set to 1800 h.

3. Results and Analysis

3.1. Results and Analysis of Bending Static Load Test

The load-displacement curves are shown in Figure 4; the mid-span displacements of the three specimens show three stages of deformation characteristics, which are elastic stage I, elastic–plastic stage II, and plastic stage III. The loads of the three specimens in the elastic stage I grew in proportion to each other at the same displacement, and the elastic stage of the GFRP-sheets-reinforced specimen was the longest. The GFRP-sheets-reinforced WPC specimen reached elastic–plastic stage II when the span displacement increased to 15 mm, showing a small decrease in load while the span displacement increases quickly, fluctuated in a small range, and then the specimen yielded and reached the elastic–plastic stage II, with a displacement of 30 mm to reach the plastic stage III. At this time, the mid-span displacement reached 44.37 mm, which was the second largest in these specimens, then shear failure occurred. The GFRP-rebar-reinforced specimen reached the elastic–plastic stage when the mid-span displacement was about 5 mm, which is almost the same as the unreinforced specimen. The displacement gradually increased during the loading process; when the load gradually became stable, the specimen had the obvious plasticity of stage III, and the displacement still increased. The GFRP-rebar-reinforced specimen had the best ductility and the largest displacement of 66.77 mm.

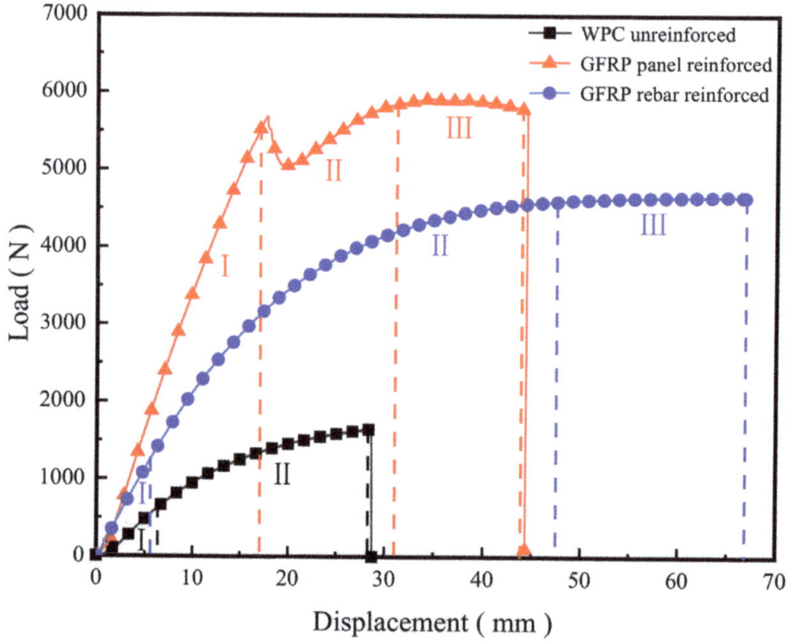

Figure 4. Load-mid-span displacement curves of WPC panels in static load test.

The results of the three-point bending loading test are shown in Table 1. The bearing capacity and ductility of the two GFRP-reinforced specimens are greatly increased. Compared with the unreinforced specimen, the average ultimate bearing capacity of the GFRP-sheets-reinforced specimens was increased by 258% and the mid-span displacement was increased by 55%; the average ultimate bearing capacity of the GFRP-rebar-reinforced specimens was increased by 165% and the mid-span displacement was increased by 133%.

Table 1. Static load test results of three-point bending for WPCs.

Specimen Types	Failure Mode	Specimen Number	Ultimate Load (N)	The Average (N)	Mean Mid-Span Displacement (mm)	The Average (mm)
Unreinforced	Shear failure	1 2 3	1738 1715 1707	1720	27.62 28.46 29.81	28.63
Reinforced with GFRP sheet	Bending shear failure	1 2 3	6165 6151 6134	6150	43.24 44.12 45.75	44.30
Reinforced with GFRP rebar	Exceed displacement limit	1 2 3	4576 4558 4546	4560	65.22 65.86 69.23	66.70

3.2. *Results and Analysis of Bending Creep Test*

3.2.1. Failure Mode

During the bending creep test, failure did not occur under low load levels. The mid-span deflection of the specimen increased significantly during the transient loading, and with the increase in time, the creep deflection of each specimen increased slowly and stabilized, which showed two stages of typical creep, transient creep and deceleration creep, and none of them reached the accelerated creep damage stage. Shear failure occurred in unreinforced specimens at the 80% load level. At the instant of loading, the mid-span deflection of the specimen increased rapidly accompanied by a slight resin fracture sound. With the loading time increasing, many small cracks appeared on the lower surface of the specimens after 1 h, the cracks kept extending upward, and they finally fractured after 7 h, which is a typical brittle failure, as shown in Figure 5a. The movement of molecular chains mainly occurs in the creep process of polymer materials. In the transient creep phase, the bond lengths and bond angles of molecular chains are changed due to the loading effect, and the curled chains are unfolded. When the unfolded chains are restrained, the deceleration creep phase is entered and the slip between different molecular chains occurs. The polymer matrix plays the role of bonding wood fibers and transferring stress inside the wood-plastic material, while the polymer molecular chain movement is restrained to some extent by the wood fibers. The GFRP-sheets-reinforced specimens made a slight sound of resin and fiber fracture during loading at the 80% load level, the GFRP sheets were folded and whitened, and the deflection deformation in the span increased significantly. A small crack appeared on the specimen after 840 h from the bottom across the middle and gradually extended, and the glass fibers were pulled off and bending shear damage occurred after 960 h; the crack suddenly appeared as shown in Figure 5b. The GFRP-rebar-reinforced specimens were considered to fail due to excessive deformation of their mid-span deflection at the higher load levels of 70% and 80%; there were no cracks or fractures, and the mid-span deflection increased significantly, as shown in Figure 5c. It can be seen that the shear resistance of the WPC specimens reinforced by GFRP sheets was improved, and the ultimate loading capacity of the specimens was significantly increased. The bending resistance of the WPC specimens reinforced by the GFRP rebar was improved, which showed better ductility and interfacial properties. The failure mode was also changed from brittle shear failure to bending shear damage or excessive mid-span deformation.

3.2.2. Creep Strain–Time Curves

Figure 6 shows the bending creep strain–time curves of WPC panels. It can be seen from Figure 6a that the strains of the unreinforced specimens gradually increased with time. The final strain rates grow by 226.67%, 182.11%, 99.61%, 88.74%, and 44.31% under the five stress levels from 30% to 70%, respectively. From Figure 6b, the final strain growth rates of the GFRP-sheet-reinforced specimens are 259.77%, 134.00%, 37.96%, 22.70%, 22.26%, and

15.25% from 30% to 80%. From Figure 6c, the final strain growth rates of the GFRP-rebar-reinforced specimens are 440.00%, 194.12%, 86.97%, 49.21%, 22.11%, and 24.66% from 30% to 80%. The initial creep strains of the three types of specimens are transient and similar. Transient deformation occurred when the designed load was carried out at different load levels, and the initial strains increased with the increase in the load level. The development trend of the initial strains of the specimens at the same load level is similar. It can be seen from the figures that the creep strains of the specimens under low load levels are relatively close, the creep deformation increases with time, the deformation rate declines gradually, and the overall creep development is relatively smooth. The creep strain of the specimens at high load levels grows rapidly and the deformation rate is significantly larger than that at low load levels.

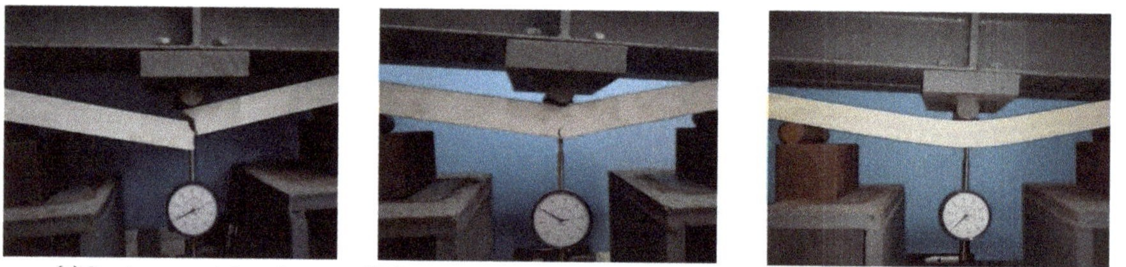

(**a**) Specimen unreinforced (**b**) Specimen reinforced with GFRP sheets (**c**) Specimen reinforced with GFRP rebar

Figure 5. Failure mode of PVC-based WPC specimens.

As shown in Figure 6a, after the initial quick increasing stage, the creep strain of the unreinforced specimens gradually increases with time and the strain rate shows a deceleration trend. With the increase in the load level, the strain rate increases. Shear failure occurred rapidly after 7 h loading at the 80% load level. The time is so short that we cannot see this in Figure 6a, while the specimens under other load levels reach the decelerating creep stage and stabilize after the transient creep stage.

The strain of the GFRP-sheets-reinforced specimens is shown in Figure 6b; cracks failure occurred after 1000 h at the 80% load level, which is in the creep stabilization stage. The strain almost kept stable with time after 1200 h at the lower load level. Figure 6c represents that the initial strains of the GFRP-rebar-reinforced specimens are largest at the 80% load level, the failure did not occur but the strain rate increases significantly with time, and the strain should not be negligible at the last stage.

Figure 7 shows the bending creep strain–time curves of the three types of specimens at the same loads. As shown in Figure 7a, the bending creep curves of unreinforced and GFRP-reinforced specimens at the same load of 1400 N are significantly different; the unreinforced specimen shows a significant increase in strain during transient loading and a non-stationary rapidly increasing strain rate during 7 h loading. The strain of GFRP-rebar-reinforced specimens increased during the transient loading and gradually increased during the loading process, the strain rate gradually decreased and became stable, and creep development was relatively smooth overall. The final strain value of the unreinforced specimen was 960% higher than that of the GFRP-rebar-reinforced specimen; the GFRP-rebar-reinforced method not only improved the load capacity of WPCs but also reduced the creep strain effectively.

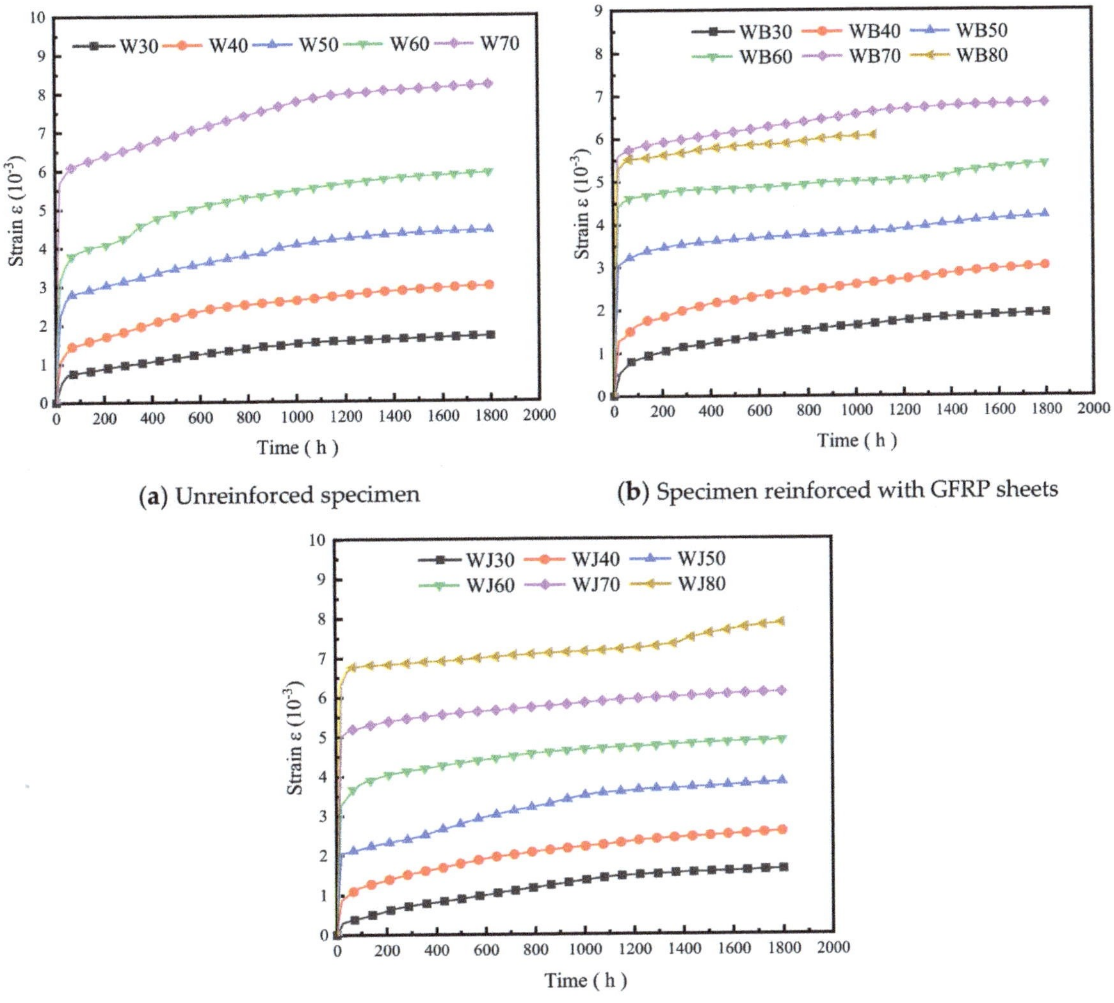

(a) Unreinforced specimen

(b) Specimen reinforced with GFRP sheets

(c) Specimen reinforced with GFRP rebar

Figure 6. Bending creep strain–time curve of PVC-based WPCs (W represents the unreinforced WPC specimen type, WB represents the type reinforced with GFRP sheets, and WJ represents the type reinforced with GFRP rebar).

Figure 7b demonstrates the bending creep curves of GFRP-sheets- and rebar-reinforced specimens at the same load of 3600 N; the change trend is similar in shape, both strains increased rapidly during transient loading and the strain tended to be stable after 72 h, the strain rate gradually decreased, and the overall creep strain development was relatively smooth. The final strain value of GFRP-rebar-reinforced specimens was 150% higher than that of the GFRP-sheets-reinforced specimens. With the increase in time, the strain value of the GFRP-sheets-reinforced specimens was always smaller than that of GFRP-rebar-reinforced specimens, which indicated that the former had stronger bending resistance and a better reinforcement effect than the latter.

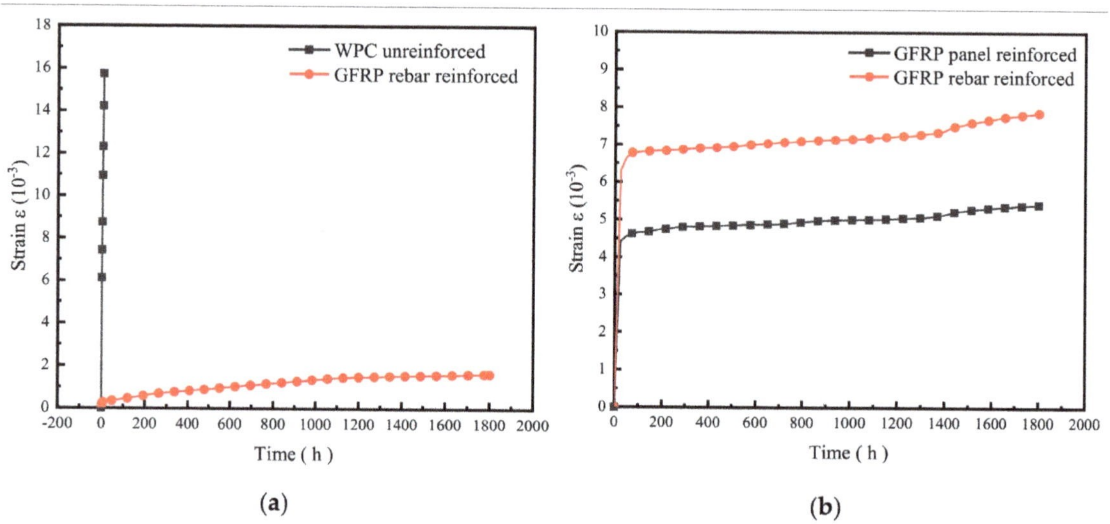

Figure 7. Comparison of bending creep strain–time curves of unreinforced and reinforced WPCs at the same load. (**a**) Strain comparison between unreinforced and reinforced with GFRP rebar specimens (load 1400 N). (**b**) Strain comparison between GFRP-sheets- and rebar-reinforced specimens (load 3600 N).

Overall, at the same load, the creep resistance of the GFRP-sheets-reinforced specimen is the strongest and the reinforcement effect is the best, and the creep resistance of the GFRP-rebar-reinforced specimens is much better than that of the unreinforced specimens. The best reinforcement method to reduce bending creep deformation is reinforcement with GFRP sheets.

In order to study the growth law of creep strain with time variation at different load levels and to evaluate the variation law of creep more reasonably, the definition of relative creep from ASTM D6815-09 [65] is:

$$\zeta = \varepsilon_t / \varepsilon_o \quad (1)$$

where ζ is the relative creep of the strain at time t relative to the initial strain; ε_t is the strain at time t; and ε_o is the initial strain.

The test data were processed uniformly to obtain the relative creep–time curves of the three specimen types, as shown in Figure 8. The curves show that the value of relative creep increases with time. The final relative creep of the unreinforced specimens was 327%, 282%, 199%, 189%, and 144% of the initial values under each load level from 30% to 70%; the GFRP-sheets-reinforced specimens were 360%, 234%, 138%, 123%, 122%, and 121% of the initial values from 30% to 80%; and the GFRP-rebar-reinforced specimens were 540%, 294%, 187%, 149%, 122%, and 122% of the initial values from 30% to 80%. The relative creep growth rate of the three specimen types was generally decelerated, indicating that the second stage of creep of wood-plastic specimens had a decelerated creep stage. Since the initial strain value is larger for specimens with a high load level under transient loading, the strain value changes gently during the deceleration creep phase, so the relative creep decreases with the increase in load level. The slope of the curve differs for specimens with different load levels. For the specimens of the same type, the higher the load level, the smaller the relative creep value and the slower the growth at a later stage. The relative creep variation trends of the three types of members are closer when the load rating is higher.

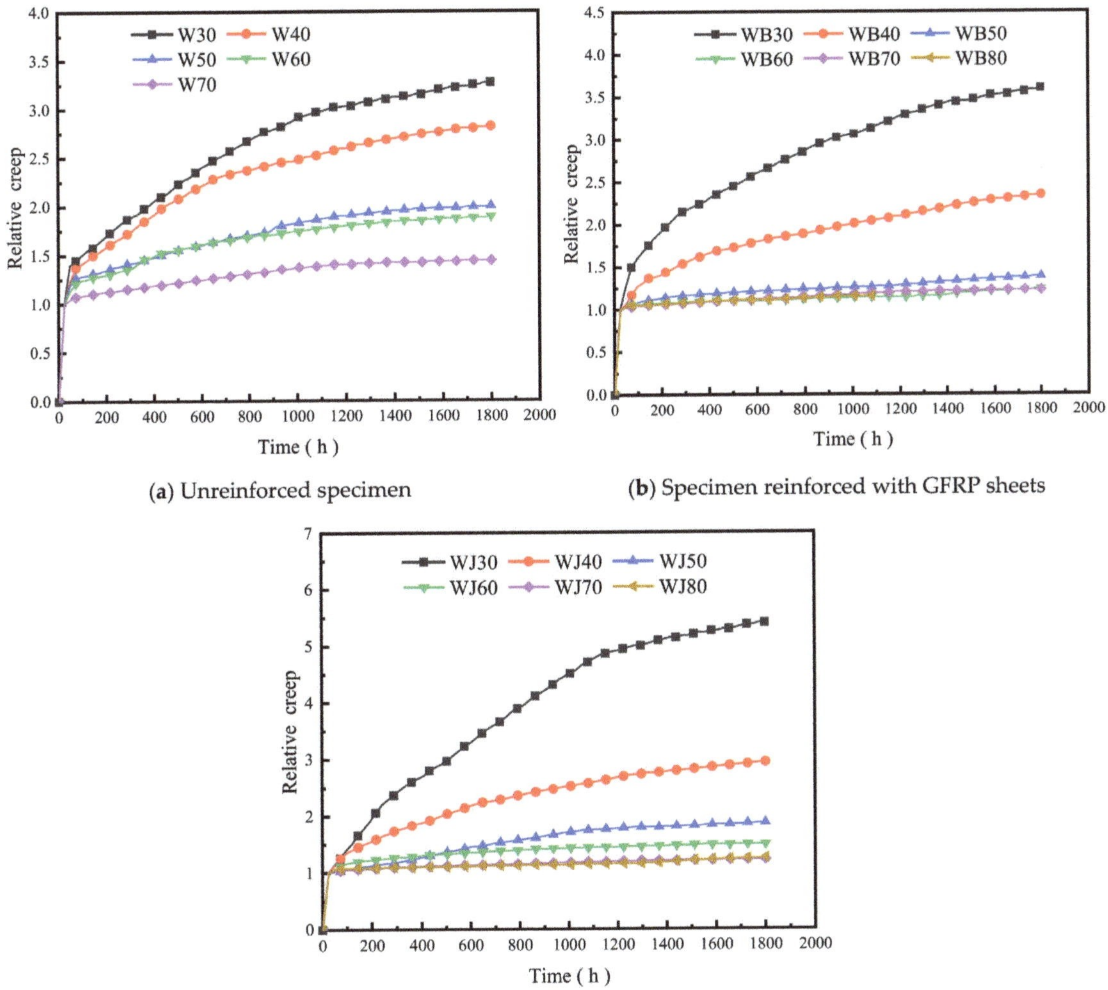

(**a**) Unreinforced specimen

(**b**) Specimen reinforced with GFRP sheets

(**c**) Specimen reinforced with GFRP rebar

Figure 8. Relative creep–time curve of PVC-based WPCs.

4. Four-Element Theoretical Model

Based on the above experiment analysis, the bending creep behavior of the specimen is consistent with the four-component model, which describes the development law of viscoelastic creep, which consists of one spring unit, one Kelvin–Voigt model, and one viscous pot unit in series, as shown in Figure 9, which is used to describe the deformation characteristics of the three parts of the material: normal elastic deformation, high elastic deformation, and viscous flow, respectively.

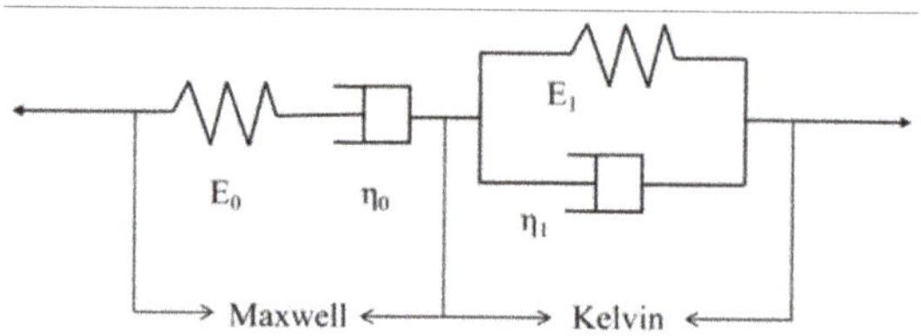

Figure 9. Four-element model.

The generic expression for the model is:

$$\varepsilon(\mathrm{t}) = \frac{\sigma}{E_0} + \frac{\sigma_t}{\eta_0} + \frac{\sigma}{E_1}[1 - \exp(-tE_1/\eta_1)] \quad (2)$$

where t is time; $\varepsilon(t)$ is the total material strain at time t; σ is the constant stress; E_0 is the material modulus of elasticity; η_0 is the material viscosity coefficient; E_1 is the elasticity coefficient; and η_1 is the viscosity coefficient.

The model equation can represent the three components of material creep: σ/E_0 for elastic deformation, σ_t/η_0 for viscous deformation, and $(\sigma/E_1)[1 - \exp(-tE_1/\eta_1)]$ for viscoelastic deformation. It is known from the deformation element that the viscous creep of the model increases with time if the viscosity coefficient η_1 is determined. The model was used to fit and analyze the bending creep test results of PVC wood-plastic composites to obtain the creep strain–time curve, as shown in Figure 10.

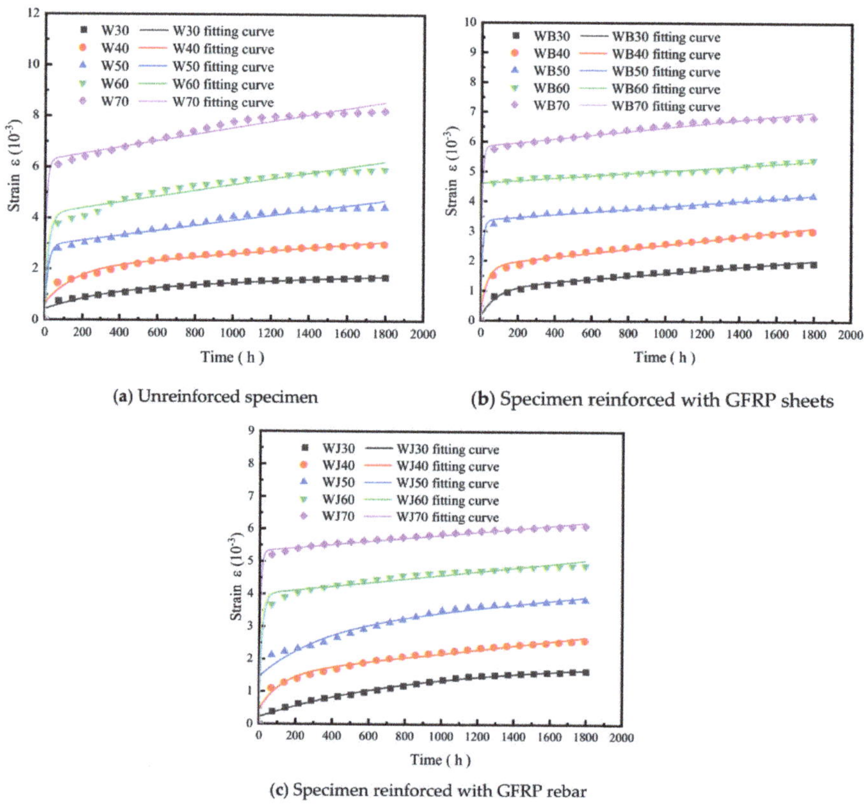

Figure 10. Fitting curve of four-element model of PVC-based WPCs.

Figure 10 shows that the fitted curves of the four-element model are closer to the test results under low load levels and the correlation coefficients of the curve fits are larger than 0.97, with good fitting accuracy. The fitting error of unreinforced specimens increases slightly and the strain rate increases as the load level increases; the correlation coefficients of the fitted curves for both GFRP-sheets- and rebar-reinforced specimens are high for each load level, which shows that the four-element model is suitable for predictive fitting analysis of creep strain in two stages of transient creep and deceleration creep of WPCs and can be used for long-term creep fitting analysis and life prediction of creep in the first two stages.

5. Conclusions

GFRP-sheets- and GFRP-rebar-reinforced specimens were designed to improve the bearing capacity and reduce the bending creep performance of WPCs, the static experiments and long-time bending creep experiments were carried out, and creep strain prediction analysis was accomplished based on the four-element model. The conclusions are as follows:

(1) The ultimate load-bearing capacity of WPC panels is greatly improved with GFRP sheets and rebar reinforcement, and the ultimate bearing capacity of the GFRP-sheets-reinforced specimen increased by 258% and the GFRP-rebar-reinforced specimen increased by 165%. Using GFRP to reinforce WPCs is effective.

(2) The failure mode of PVC-based WPCs is closely related to the load level and the reinforcement method. After the transient stage, the creep kept decreasing and developed stably without damage under the low load level. When the loading level reached 80% of the ultimate load, unreinforced specimens suffered shear failure after 7 h, the maximum deflection in the span was 19.10 mm. The mid-span deflection of the GFRP-sheets-reinforced specimen increased significantly under transient loading, the GFRP sheets' fibers were pulled out and bending shear failure occurred after 960 h, and the maximum deflection in the span was 7.03 mm. GFRP-rebar-reinforced specimens had significant deformation and were considered to fail due to their excessive mid-span deflection deformation, no cracks or fractures appeared, and the specimens had excellent ductility.

(3) The creep strain change rules of specimens were obtained under different load levels. The strain increased and the strain growth rate decreased with time. The final strain rates of the unreinforced specimen increased by 226.67%, 182.11%, 99.61%, 88.74%, and 44.31% under the loading levels from 30% to 70%; those of the GFRP-sheets-reinforced specimen were 259.77%, 134.00%, 37.96%, 22.70%, 22.26%, and 15.25% under the loading levels from 30% to 80%; those of the GFRP-rebar-reinforced specimen were 440.00%, 194.12%, 86.97%, 49.21%, 22.11%, and 24.66% under the loading levels from 30% to 80%. The strain growth rates of the three types of specimens are generally decelerated. The relative creep decreased with increasing load levels. The final relative creep of unreinforced specimens was 327%, 282%, 199%, 189%, and 144% under the load levels from 30% to 70%; that of the GFRP-sheets-reinforced specimens was 360%, 234%, 138%, 123%, 122%, and 121%; and that of the GFRP-rebar-reinforced specimens was 540%, 294%, 187%, 149%, 122%, and 122% under the load levels from 30% to 80%.

(4) Based on the test results, a four-element model was built. It is a simple, accurate, and reasonable classical model for describing the creep performance of WPC either reinforced by GFRP or not. The model can be used for long-term strain analysis and life prediction of specimens in the first two stages of transient creep and deceleration creep. GFRP-reinforced design methods can effectively improve the mechanical properties and creep resistance of WPCs and broaden the application field of WPC products.

Author Contributions: B.D.: Data curation and analysis, Writing—original draft. R.H.: Conceptualization, Writing—review, Validation review, Supervision, Funding acquisition. K.W.: Investigation, Writing—editing, Visualization. Z.M.: Investigation, Formal analysis. H.F.: Methodology, Project administration, Resources. All authors have read and agreed to the published version of the manuscript.

Funding: This research was funded by the National Key Research and Development Program of China (Grant No. 2019YFD1101205).

Institutional Review Board Statement: Not applicable.

Informed Consent Statement: Not applicable.

Data Availability Statement: The data presented in this study are available on request from the corresponding author.

Conflicts of Interest: The authors declare no conflict of interest.

References

1. Alrubaie, M.A.A.; Lopez-Anido, R.A.; Gardner, D.J. Flexural creep behavior of high-density polyethylene lumber and wood plastic composite lumber made from thermally modified wood. *Polymers* **2020**, *12*, 262. [CrossRef] [PubMed]
2. Abdullah, A.; Emin, S.; Hayrettin, D.; Ömer, S.S.; Mürsel, E. The effects of harsh aging environments on the properties of neat and MWCNT reinforced epoxy resins. *Constr. Build. Mater.* **2021**, *272*, 121929.
3. Abouelregal, A.E.; Marin, M. The size-dependent thermoelastic vibrations of nanobeams subjected to harmonic excitation and rectified sine wave heating. *Mathematics* **2020**, *8*, 1128. [CrossRef]
4. Wang, Q.W.; Wang, W.H. *Wood-Plastic Composites and Products*; Chemical Industry Press: Beijing, China, 2007.
5. Wang, D.Z.; Zhu, M. Research progress of biomass WPCs. *Sci. Technol. Innov.* **2019**, *22*.
6. Li, C.; Xian, G.; Li, H. Effect of postcuring immersed in water under hydraulic pressure on fatigue performance of large- diameter pultruded carbon/glass hybrid rod. *Fatigue Fract. Eng. Mater.* **2019**, *42*, 1148–1160. [CrossRef]
7. Lal, H.M.; Uthaman, A.; Li, C.G.; Xian, G.J.; Thomas, S. Combined effects of cyclic/sustained bending loading and water immersion on the interface shear strength of carbon/glass fiber reinforced polymer hybrid rods for bridge cable. *Constr. Build. Mater.* **2022**, *314*, 125587. [CrossRef]
8. Pan, Y.F.; Yan, D.M. Study on the durability of GFRP bars and carbon/glass hybrid fiber reinforced polymer (HFRP) bars aged in alkaline solution. *Compos. Struct.* **2021**, *261*, 113285. [CrossRef]
9. Bledzki, K.; Andrzej; Omar, F. Creep and impact properties of wood fibre-polypropylene composites: Influence of temperature and moisture content composites. *Sci. Technol.* **2004**, *64*, 693–700. [CrossRef]
10. Adrian, J.; Marcovich, E. Norma Analysis of the Creep Behavior of Poly propylene-Woodfiber Composites. *Polym. Eng. Sci.* **2004**, *44*, 1594–1603.
11. Lee, S.Y.; Yang, H.S.; Kim, H.J.; Jeong, C.S.; Lim, B.S.; Lee, J.N. Creep behavior and manufacturing parameters of wood fiber filled poly propylene composites. *Compos. Struct.* **2004**, *65*, 459–469. [CrossRef]
12. Hung, K.C.; Wu, L.T.; Chen, Y.L. Assessing the effect of wood acetylation on mechanical properties and extended creep behavior of wood/recycled-polypropylene composites. *Constr. Build. Mater.* **2016**, *108*, 139–145. [CrossRef]
13. Pooler, D.J.; Smith, L.V. Nonlinear Viscoelastic Response of a Wood- Plastic Composite Including Temperature Effects. *J. Thermoplast. Compos.* **2004**, *17*, 425–450. [CrossRef]
14. Park, P.D.; Balatinecz, J.J. Short term flexural creep behavior of wood- fiber/polypropylene composites. *Polym. Compos.* **2004**, *19*, 377–382. [CrossRef]
15. Fang, H.; Sun, H.; Liu, W.; Wang, L.; Bai, Y.; Hui, D. Mechanical performance of innovative GFRP-bamboo-wood sandwich beams: Experimental and modelling investigation. *Compos. Part B Eng.* **2015**, *79*, 182–196. [CrossRef]
16. Tuwair, H.; Hopkins, M.; Volz, J.; Eigawady, M.A.; Mohamed, M.; Chandrashekhara, K.; Birman, V. Evaluation of sandwich panels with various polyurethane foam-cores and ribs. *Compos. Part B Eng.* **2015**, *79*, 262–276. [CrossRef]
17. Shi, H.; Liu, W.; Fang, H.; Bai, Y.; Hui, D. Flexural responses and pseudo-ductile performance of lattice-web reinforced GFRP-wood sandwich beams. *Compos. Part B Eng.* **2017**, *108*, 364–376. [CrossRef]
18. Gigliotti, L.; Pinho, S.T. Prediction of the post-crushing compressive response of progressively crushable sandwich foam cores. *Compos. Part A Appl. Sci.* **2016**, *80*, 148–158. [CrossRef]
19. Zhang, P.; Cheng, Y.; Liu, J.; Li, Y.; Zhang, C.; Hou, H.; Wang, C. Experimental study on the dynamic response of foam-fifilled corrugated core sandwich panels subjected to air blast loading. *Compos. Part B Eng.* **2016**, *105*, 67–81. [CrossRef]
20. Chen, Q.; Linghu, T.; Gao, Y.; Wang, Z.; Liu, Y.; Du, R.; Zhao, G. Mechanical properties in glass fiber PVC-foam sandwich structures from different chopped fiber interfacial reinforcement through vacuum-assisted resin transfer molding (VARTM) processing. *Compos. Sci. Technol.* **2017**, *144*, 202–207. [CrossRef]
21. Badaruzzaman, W.H.W.; Dabbagh, N.M.R.; Salleh, K.M.; Saharuddin, E.N.; Radzi, N.F.M.; Azham, M.A.A.; Sani, S.F.A.; Zakaria, S. Mechanical Properties and Water Absorption Capacity of Hybrid GFRP Composites. *Polymers* **2022**, *14*, 1394. [CrossRef]
22. Zhou, Z.L.; Yang, Y.D. A preliminary discussion of creep mechanism of fiber reinforced plastics. *FRP/Compos* **1985**, *4*, 29–33.

23. Ghita, C.; Pop, N.; Popescu, I.N. Existence result of an effective stress for an isotropic visco-plastic composite. *Comp. Mater. Sci.* **2012**, *64*, 52–56. [CrossRef]
24. Bai, X.Y. *Experimental Study and Theoretical Analysis of Anchoring Mechanism of GFRP Anti-Floating Anchors*; Qingdao University of Technology: Qingdao, China, 2015.
25. Mu, X.Y. *Creeping Mechanics*; Xi'an Jiaotong University Press: Xi'an, China, 1990.
26. Shen, S.Z. Creep properties of FRP. *Q. J. Mech.* **1982**, *2*, 71–78.
27. Dutta, P.K.; Hui, D. Creep rupture of a GFRP composite at elevated temperatures. *Comput. Struct.* **2000**, *76*, 153–161. [CrossRef]
28. Tannous, F.E.; Saadatmanesh, H. Durability of AR glass fiber reinforced plastic bars. *J. Compos. Constr.* **1999**, *3*, 12–19. [CrossRef]
29. Uomoto, T.; Nishimura, T. Deterioration of aramid, glass, and carbon fibers due to alkali, acid, and water in different temperatures. *ACI Spec. Publ.* **1999**, *188*, 515–522.
30. Li, G.W.; Gao, L.; Huang, Z.H.; Liu, C.Q.; Zhang, D. Experiments on the pull-out model of damage mechanism of full-length bonded glass fiber reinforced polymer anchors. *J. Rock. Mech. Geotech.* **2007**, *26*, 1653–1663.
31. Li, G.W.; Liu, Z.Q.; Huang, Z.H.; Cai, Y.Q. Field test of applying glass fiber anchors to strengthen highway slopes. *J. Rock. Mech. Eng.* **2010**, *29* (Suppl. 2), 4056–4062.
32. Sun, Q. *Experimental Study on the Mechanical Properties of GFRP Anchors under Freeze-Thaw Cycles and Long-Term Loading*; Central South University: Changsha, China, 2012.
33. Cui, Y.H.; Noruziaan, B.; Lee, S.; Chang, M.; Tao, J. Glass fiber/wood plastic hybrid composites and their synergistic reinforcing effects. *Polym. Mater. Sci. Eng.* **2006**, *22*, 231–234.
34. Guo, D.; Lu, S.R.; Luo, C.X.; Liu, K.; Ling, R.H. Study on the structure and properties of glass fiber reinforced lignin/PP composites. *Modern Plast. Process. Appl.* **2012**, *24*, 23–26.
35. Jeamtrakull, S.; Kositchaiyong, A.; Markpin, T.; Rosarpitak, V.; Sombatsompop, N. Effects of wood constituents and content, and glass fiber reinforcement on wear behavior of wood/PVC composites. *Compos. Part B Eng.* **2012**, *43*, 2721–2729. [CrossRef]
36. Huang, R.; Xiong, W.; Xu, X.; Wu, Q. Thermal expansion behavior of co-extruded WPCs with glass-fiber reinforced shells. *Bioresources* **2012**, *7*, 5514–5526. [CrossRef]
37. Kim, B.J.; Huang, R.; Han, J. Mechanical and morphological properties of coextruded wood plastic composites with glass fiber-filled shell. *Polym. Compos.* **2014**, *37*, 824–834. [CrossRef]
38. Zolfaghari, A.; Behravesh, A.H.; Adli, A. Continuous glass fiber reinforced wood plastic composite in extrusion process: Mechanical properties. *Mater. Des.* **2013**, *51*, 701–708. [CrossRef]
39. Pulngern, T.; Udtaranakron, T.; Chanto, K. Physical and Mechanical Behaviors of Thermally Modified Rubberwood Glulam Beam Under Sustained and Cyclic Loading. *Wood Fiber Sci.* **2020**, *52*, 1–15. [CrossRef]
40. Pulngern, T.; Chitsamran, T.; Chucheepsakul, S.; Rosarpitak, V.; Patcharaphun, S.; Sombatsompop, N. Effect of temperature on mechanical properties and creep responses for wood/PVC composites. *Constr. Build. Mater.* **2016**, *111*, 191–198. [CrossRef]
41. Sain, M.M.; Balatinecz, J.; Law, S. Creep fatigue in engineered wood fiber and plastic compositions. *J. Appl. Polym. Sci.* **2000**, *77*, 260–268. [CrossRef]
42. Li, X.L.; Liu, W.Q.; Fang, H.; Huo, R.L.; Wu, P. Flexural creep behavior and life prediction of GFRP-balsa sandwich beams. *Compos. Struct.* **2019**, *224*, 111009. [CrossRef]
43. Tian, X.L.; Li, Z.G.; Jiang, Y.T.; He, Q.; Wu, C.Y. Study of creep properties of WPCs under different loading methods. *Plast. Ind.* **2008**, *36*, 43–46.
44. Cao, Y.; Xu, H.L.; Wang, W.H.; Wang, Q.W. Creep properties and creep modeling of molded poplar fiber/high density polyethylene composites. *J. Compos.* **2016**, *33*, 1174–1178.
45. Du, H.H.; Wang, W.H.; Wang, H.G.; Wang, Q.W. Effect of wood fiber content on creep properties of WPCs. *J. Constr. Mater.* **2015**, *18*, 333–339.
46. Xie, W.R. *Design of Creep Testing Device and Creep Performance Study of WPCs*; Northeast Forestry University: Harbin, China, 2017.
47. Cao, J.Z. *Wood Conservation and Modification*; China Forestry Press: Beijing, China, 2018.
48. Lu, X.C.; Fang, Q.H.; Lu, Q.Z.; Chen, F.Q.; Sun, X.M. Creep properties of PP/wood flour composites. *Plastics* **2009**, *38*, 81–84.
49. Jiang, Y.T.; Li, Z.G.; Wu, Z.Y.; Ding, J.S. Creep and stress relaxation properties of WPCs. *For. Mach. Woodwork. Equip.* **2009**, *37*, 24–26.
50. Shao, X.; He, C.X.; Jiang, C.Y. Creep and thermal stability of wood fiber/PVC composites. *J. Mater. Sci. Eng.* **2019**, *37*.
51. Xu, H.L.; Cao, Y.; Wang, W.H.; Wang, Q.W.; Wang, H.G. Effect of poplar fiber size on mechanical and creep properties of hot compression molded poplar fiber/high density polyethylene composites. *J. Compos.* **2016**, *33*, 1168–1173.
52. Zhang, X.L.; Deng, Z.C. Effects of Seawater Environment on the Degradation of GFRP Composites by Molecular Dynamics Method. *Polymers* **2022**, *14*, 2804. [CrossRef]
53. Corigliano, A.; Rizzi, E.; Papa, E. Experimental characterization and numerical simulations of a syntactic-foam/glass-fibre composite sandwich. *Compos. Sci. Technol.* **2000**, *60*, 2169–2180. [CrossRef]
54. Daniel, I.M.; Gdoutos, E.E. Failure Modes of Composite Sandwich Beams. *Theor. Appl. Mech.* **2009**, *35*, 197.
55. George, J.; Sreekala, M.; Thomas, S. A review on interface modification and characterization of natural fiber reinforced plastic composites. *Polym. Eng. Sci.* **2001**, *41*, 1471–1485. [CrossRef]
56. Sheng, G.Y.; Feng, M.B. *Engineering Materials Testing Technology*; China Metrology Press: Beijing, China, 2007.

57. Hu, J.H.; Li, Y.P.; Chen, W.J.; Yang, D.Q. A combined loading-creep model of ETFE foils for flat-patterning structures. *Thin. Wall. Struct.* **2020**, *157*, 106976. [CrossRef]
58. Chen, B.; Guo, L.P.; Zhang, W.X.; Bai, Y.; Wang, X.F. Compressive creep behavior of cellulose fiber reinforced concrete. *IOP Conf. Ser. Earth Environ. Sci.* **2021**, *825*, 012022.
59. Song, Y.Q.; Li, X.L.; Ma, H.F.; Fu, H. Exploration of the full process of creep in generalized Kelvin model based on damaged body elements. *Appl. Math. Mech.* **2021**, *42*, 637–644.
60. Shi, T.B.; Hu, J.H.; Chen, W.J.; Gao, C.J. A refined numerical model for determining inflation-burst behavior of composite membrane structures. *Polym. Test.* **2020**, *81*, 106123. [CrossRef]
61. Ye, L.P.; Feng, P. Application and development of FRP in engineering structures. *Chin. Civil. Eng. J.* **2006**, *39*, 24–36.
62. Li, C.G.; Guo, R.; Wang, J.Q.; Huang, H.X.; Xian, G.J. Performance evolution of CFRP@GFRP hybrid composite rods under water immersion environment. *J. Compos. Mater.* **2021**, *38*, 3290–3301.
63. Feng, P. Development and application of composite materials in civil engineering. *FRP/Compos.* **2014**, *9*, 99–104.
64. Cui, X.; Liu, J.; Xiao, J.Y.; Zhong, Y. Research progress of vacuum introduction molding process. *Mater. Guide* **2013**, *27*, 14–18.
65. *ASTM D6815−09*; Standard Specification for Evaluation of Duration of Load and Creep Effects of Wood and Wood-Based Products. ASTM International: West Conshohocken, PA, USA, 2015.

polymers

MDPI

Article

New Self-Repairing System for Brittle Matrix Composites Using Corrosion-Induced Intelligent Fiber

Yuyan Sun [1], Dongkai Wang [1], Zuquan Jin [2], Jianwei Sun [1] and Ziguo Wang [1,*]

1 School of Civil Engineering, Qingdao University of Technology, Qingdao 266520, China
2 Engineering Research Center of Concrete Technology in Marine Environment, Ministry of Education, Qingdao University of Technology, Qingdao 266520, China
* Correspondence: wangziguo@qut.edu.cn or wangziguo2008@163.com

Abstract: Brittle matrix composites such as concrete are susceptible to damage in the form of cracks. Most of the current self-repair and self-healing techniques have repair limits on crack widths or high costs of an external stimulator, or have an unfavorable effect on the composite's strength. This paper proposes a new concept of corrosion-induced intelligent fiber (CIF) and a new self-repairing system that uses the CIFs to close cracks in brittle matrix composites within a corrosive environment without external help, and without compromising the strength. The CIF comprises an inner core fiber and an outer corrodible coating that are in equilibrium, with the core fiber in tension and the corrodible coating in compression. The preparation steps and shape recovery mechanism of the CIF and the self-repair mechanism of the CIF composites are explained. Based on these concepts, this paper also describes several mechanical models built to predict the magnitude of pre-stress stored in the core fiber, and the maximum pre-stress released to the matrix composites, and the minimum length of the reliable anchor ends of CIF. The sample calculation results show that the recovery strain was 0.5% for the CIF with the steel core fiber and 12.7% for the CIF with the nylon core fiber; the maximum crack closing force provided by the CIF to concrete can be increased by increasing the amount of the CIFs in concrete and the initial tensile stress of the core fiber. This paper provides some suggestions for enhancing the self-repair capability of brittle composites in complex working environments.

Keywords: brittle matrix composites; concrete; corrosion; crack; intelligent fiber; self-repair

1. Introduction

Brittle matrix composites such as ceramics, concrete, and brick are susceptible to cracking due to their low tensile strength, which can affect the overall mechanical performance and durability of the structures made of the composites [1–3]. Traditional repairing methods such as timed repair and after-the-fact repair can usually manage the visible cracks on the surface of some structural elements, but can hardly reach some cracks in complex structures. Self-repair or self-healing functionality of the composites, however, can provide timely repair to the structures and minimize the negative effect of cracks [3–5]—it is of great significance especially for the structures in complex environments such as space capsules, nuclear power plants, and marine tunnels.

Concrete is one of the most widely used construction materials. It is well known that concrete has some ability to self-heal [6]; its microdamage and microcracks caused by loads or environmental factors during the service time can be self-healed to some extent by further hydration of the unhydrated particles [7]. However, the self-healing efficiency of concrete seems to be low and the effect is limited by crack widths up to 200 μm [8–10]. To promote the self-repair or self-healing capacity of concrete, many techniques have been developed over the years, including: (1) Chemical self-healing techniques, such as mixing crystalline admixtures into concrete or brushing a layer of active admixture coating on the concrete surface [11–17], after water penetrates concrete through cracks, the active functional group undergoes a condensation reaction and produces $CaCO_3$ crystals that fill the capillaries and microcracks in the concrete. When the concrete is again cracked, the active molecules are reactivated and continue to react until the cracks are healed. However, this method has minimal effect on repairing cracks over 400 μm in width [18–20]. (2) There are physical self-repair techniques, such as embedding shape memory alloy (SMA) into the crack-prone parts of concrete [21–30]; when cracks are generated, the electrical heating method is used to stimulate the shape recovery of SMA, forcing the cracks to close. However, the application scope of this technique is somewhat limited due to the relatively high investment in the external stimulation system and the expensive SMA. (3) Physical–chemical self-repair techniques, such as mixing the microencapsulated/hollow fibers (carriers) containing repair adhesive into concrete [31–44]—when cracks pass through the carrier, the repair agent flows out and penetrates into the cracks. The effect of this method is influenced by the number of carriers: fewer carriers are not enough to fill the cracks, while more carriers reduce the strength of the concrete. (4) Microbial self-healing techniques, such as adding specific harmless bacteria such as aerobic alkalophilic Bacillus into concrete [45–56]—as the concrete cracks, the infiltration of oxygen and water activates the dormant bacterial spores, and the process of microbial metabolism produces CO_2 that reacts with Ca^{2+} in the concrete to produce $CaCO_3$ crystals that seal and repair cracks up to 500 μm wide [57–59]; however, this method has some environmental and practical limitations because the bacteria have certain requirements concerning working environment and temperature. (5) Electrochemical self-repair techniques, such as the electrochemical deposition method, are generally achieved in an electric field where the positive and negative ions in the solution are deposited in the crack through an electrode reaction, producing insoluble crystals that fill the cracks [60–70]. However, this method requires the entire repair process to be electrified and the cost is relatively high.

As mentioned above, most of the current self-repair and self-healing techniques have repair limitations on crack width, or have an unfavorable effect on concrete strength, or need extra investment in an external stimulus. In contrast, the addition of fibers into concrete, though they cannot close or repair cracks, is helpful to control and reduce cracking and usually results in improved concrete strength and toughness, so it often combines with some self-repair or self-healing technique to achieve crack closure [3,71–76]. However, for the structures in corrosive environmental conditions, once the concrete is cracked, the harmful substances can enter through the cracks and rapidly corrode the reinforcement and cause overall degradation of structural integrity; in this case, the effect of the current self-repair or self-healing methods is even more limited. Based on these issues, we hope to develop a novel and relatively economical self-repair system that can enhance the durability of concrete structures under harsh environmental conditions.

This paper proposes a new concept of corrosion-induced intelligent fiber (CIF) and a new self-repairing system that uses the CIFs to close cracks in the brittle matrix composites within a corrosive environment without external help. The CIF can be prepared by coating the surface of the pre-tensioned core fiber with a corrosion-prone coating, then the CIFs can be embedded in the crack-prone parts of the brittle matrix composites. Once cracking occurs, the corrosive media in the environment penetrate through the cracks and trigger the shrinkage of the CIFs, which in turn releases the stored pre-stress to drive the cracks in the brittle matrix composites to close. Based on this concept, this paper also builds several mechanical models of the CIF and the self-repairing system to predict the magnitude of pre-stress stored in the core fiber, and the maximum pre-stress released to the brittle matrix composite, and the minimum length of the reliable anchor ends of CIF. This paper aims to provide the conceptual and theoretical development of the new CIF that can be efficient for closing cracks with a wide width range without external help, and without compromising strength, which has the potential to realize the intelligent functions of the brittle matrix composites in their feedback to external corrosion damage.

2. Corrosion-Induced Intelligent Fiber (CIF)

2.1. General Concept

The new CIF comprises an inner core fiber and an outer corrodible coating, wherein the core fiber is a corrosion-resistant material, and the corrodible coating can be easily corroded by corrosive media in the environment. The preparation steps of the CIF are shown in Figure 1. First, the core fiber is in an unstressed state (see Figure 1a), secondly, it is pre-tensioned in an elastic range, and the tensile stress is σ_0 (see Figure 1b), thirdly, the surface of the core fiber is uniformly coated with a corrodible coating by deposition, spraying, electroplating, or a similar process when the tensile stress σ_0 remains constant and the corrodible coating is in an unstressed state (see Figure 1c), and fourthly, the tensile force is removed after the coating is completed and cured. It is assumed that the core fiber is well bonded to the corrodible coating. In the process of removal, the corrodible coating is axially compressed under the elastic recovery force of the core fiber, and the compressive stress is σ_c^p (see Figure 1d). Finally, a tensile–compressive equilibrium is established between the corrodible coating and the core fiber, with the former in compression and the latter in tension.

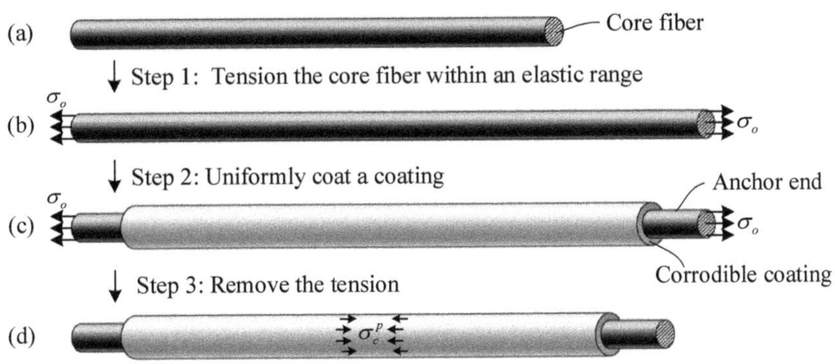

Figure 1. Schematic diagram of preparation process of the corrosion-induced intelligent fiber (CIF), (**a**) unstressed core fiber, (**b**) pre-tensioned core fiber, (**c**) pre-tensioned core fiber coated with an unstressed corrodible coating, (**d**) CIF in equilibrium with core fiber in tension and corrodible coating in compression.

2.2. Shape Recovery Mechanism of CIF

The shape recovery mechanism of the CIF is shown in Figure 2. When the environment is not corrosive, the CIF is not corroded, and the core fiber and the corrodible coating are in an original equilibrium state. In the corrosive environment, the corrodible coating in contact with the corrosive media forms a load-unbearable corrosion product, so the effective force section of the corrodible coating is decreasing, and the equilibrium state is broken; thus, the compressive stress and compressive deformation of the remaining corrodible coating increase constantly under the elastic recovery force of the core fiber, thereby the core fiber shrinks and gradually approaches its initial length, as shown in Figure 2b. Figure 2c shows that after the corrodible coating is corroded thoroughly, the pre-stress in the CIF is released and the core fiber recovers to the original length in an unstressed state.

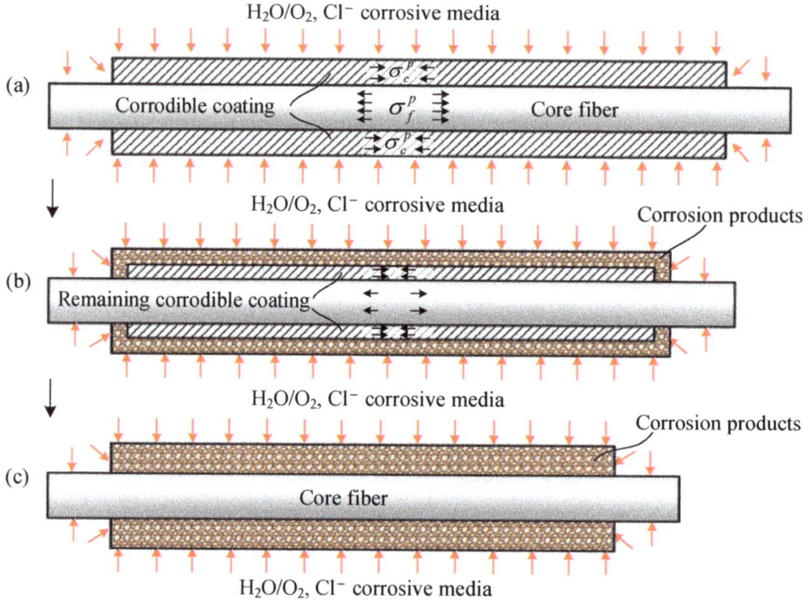

Figure 2. Axial cross-sectional view of the shape recovery mechanism of CIF, (**a**) CIF is in contact with corrosive media, (**b**) the effective force section of the corrodible coating decreases, the equilibrium is broken, and the core fiber shrinks, (**c**) core fiber recovers to the original length after the corrodible coating being corroded thoroughly.

3. Self-Repair Principle of CIF Composites

The CIF can be embedded in the crack-prone parts of the brittle matrix composite when applied in a corrosive environment. In order to transfer the pre-stress to the matrix composites more effectively, preferably the CIF reserve reliable anchor ends, such as uncoated bare ends, gradually thickening ends or end hooks. In the presence of the reliable anchor ends, whether the crack is distributed at the end portion of the CIF or the corrodible coating is completely corroded, the fiber is unlikely to be pulled out. The principle of self-repair of the CIF composites is shown in Figure 3. When the matrix composite cracks and the crack tips develop to the corrodible coating of the CIF, the corrosive media enter along the cracks and chemically or electrochemically react with the corrodible coating, and the CIF is triggered to shrink and transfers load through the bonding interface between the uncorroded CIF and the matrix composite to apply pressure to the matrix composite (see Figure 3b). It is clear that the higher corrosion degree of the corrodible coating, the larger crack closing force and a smaller crack width is generated. After the corrodible coating is corroded to a certain extent, the crack closing force is large enough, and the cracks are closed, as shown in Figure 3c; therefore, the inner passage for the corrosive media is cut off, and the corrosion is stopped, so the self-repair function is realized. At this time, the shrinkage of the CIF stops without increasing pressure to the brittle matrix composite.

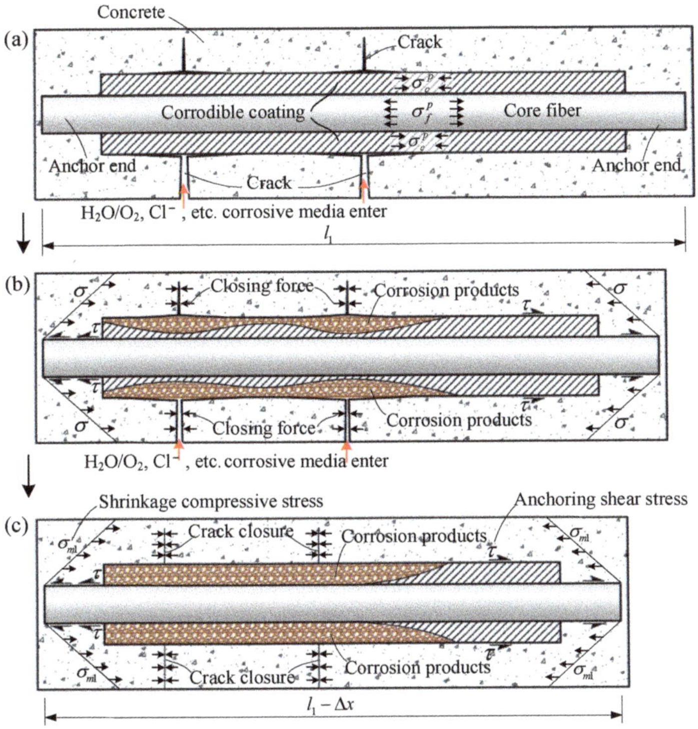

Figure 3. Axial cross-sectional view of the self-repair principle of the CIF composite, (**a**) corrosive media enter along the cracks and react with the corrodible coating, (**b**) CIF is triggered to shrink, the pre-stress stored in the core fiber is released to the matrix composite, (**c**) after the corrodible coating is corroded to a certain extent, the pressure applied to the matrix composite is large enough to close the cracks.

4. Derivation of the Mechanical Model

4.1. Mechanical Model of CIF

4.1.1. Basic Assumption

Since the CIF is a unidirectional composite with a large enough slenderness ratio, in order to simplify the calculation of internal force of the CIF, the following assumptions may be made:

1. The corrodible coating is evenly coated on the core fiber;
2. The core fiber and the corrodible coating are well bonded at the interface and the two have good chemical compatibility;
3. The influence of transverse strain of the core fiber and the corrodible coating is ignored and not incorporated into the Poisson's ratio in formula derivation;
4. The force of the core fiber and the corrodible coating is in a linear elastic state;
5. The structural unit is pulled positive and compressed negative.

4.1.2. Calculation of Internal Force

The symbols used in this section are listed in Table A1.

As shown in Figure 4, it is assumed that the original length of the core fiber that will be coated by the corrodible coating is l (see Figure 4a); in the pre-tensioning stage (see Figure 4b), the tensile stress is σ_0, and the elongation of the core fiber is Δx_1; in the coating stage (see Figure 4c), the length of the deposited coating is $l + \Delta x_1$. Owing to the recovery force of the core fiber after removing the pre-tension (see Figure 4d), the compressive deformation of the coating is Δx_2, and the core fiber and the coating achieve force equilibrium and coordinated deformation. The tensile force of the core fiber is obtained according to Hooke's law as

$$F_f = \frac{E_f A_f}{l}(\Delta x_1 - \Delta x_2) \tag{1}$$

The pressure of the corrodible coating is

$$F_c = \frac{E_c A_c}{l + \Delta x_1}(-\Delta x_2) \tag{2}$$

The force equilibrium requires $F_f + F_c = 0$, that is

$$\frac{E_f A_f}{l}(\Delta x_1 - \Delta x_2) + \frac{E_c A_c}{l + \Delta x_1}(-\Delta x_2) = 0 \tag{3}$$

then

$$\Delta x_2 = \frac{E_f A_f \Delta x_1}{\frac{E_c A_c l}{l + \Delta x_1} + E_f A_f} \tag{4}$$

Since the compressive stress in the corrodible coating is

$$\sigma_c^p = E_c \varepsilon_c = E_c \frac{-\Delta x_2}{l + \Delta x_1} \tag{5}$$

Substituting Equation (4) into Equation (5) gives

$$\sigma_c^p = -\frac{E_c E_f A_f \Delta x_1}{E_c A_c l + E_f A_f (l + \Delta x_1)} \tag{6}$$

Supposing the cross-sectional area of the CIF is $A = A_c + A_f$, and simultaneously dividing the numerator and denominator on the right side of Equation (6) by Al, then

$$\sigma_c^p = -\frac{E_c E_f V_f \varepsilon_f}{E_c V_c + E_f V_f (1 + \varepsilon_f)} \tag{7}$$

where $\varepsilon_f = \Delta x_1 / l$; substituting $\varepsilon_f = \sigma_0 / E_f$ into Equation (7), then

$$\sigma_c^p = -\frac{E_c V_f \sigma_o}{E_c V_c + E_f V_f + V_f \sigma_o} \tag{8}$$

As σ_0 is much smaller than E_f, thus,

$$\sigma_c^p \approx -\frac{E_c V_f \sigma_o}{E_c V_c + E_f V_f} \tag{9}$$

At this point, the expression of pre-stress stored in the core fiber is

$$\sigma_f^p = -\frac{\sigma_c^p V_c}{V_f} = \frac{E_c V_c \sigma_o}{E_c V_c + E_f V_f} = \frac{E_c V_c \sigma_o}{E_1} \tag{10}$$

where $E_1 = E_c V_c + E_f V_f$ is the composite elastic modulus, and $V_c + V_f = 1$.

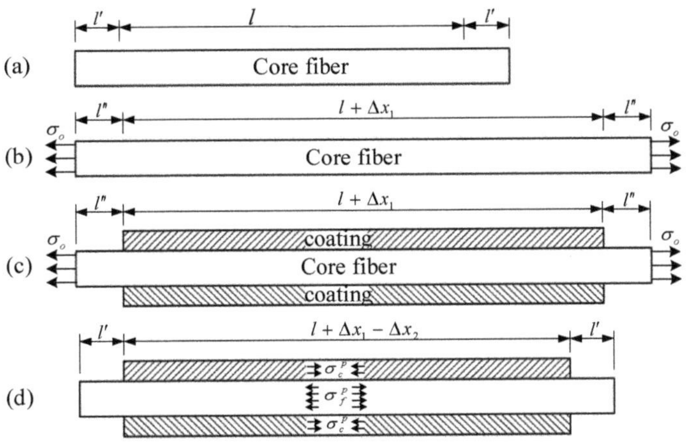

Figure 4. The process diagram of CIF force equilibrium, (**a**) unstressed stage, (**b**) pre-tension stage, (**c**) coating stage, (**d**) external force withdrawal stage.

4.1.3. Force Storage Optimization

Based on Equation (10), the axial force F stored in the core fiber is

$$F = \sigma_f^p A_f = \frac{E_c V_c \sigma_0 A_f}{E_c V_c + E_f V_f} = \frac{E_c V_c \sigma_0 V_f A}{E_c V_c + E_f V_f} = \frac{(1-V_f)V_f}{E_c(1-V_f)+E_f V_f} E_c \sigma_0 A \tag{11}$$

When F is maximum, the pre-stress released to the matrix composite is maximum. To solve the maximum value of the axial force of the core fiber, the F is first derived to obtain

$$F' = \frac{(1-2V_f)\left[E_c(1-V_f)+E_f V_f\right] - (V_f - V_f^2)(E_f - E_c)}{\left[E_c(1-V_f)+E_f V_f\right]^2} E_c \sigma_0 A \tag{12}$$

that is

$$F' = \frac{(E_c - E_f)V_f^2 - 2E_c V_f + E_c}{\left[E_c(1-V_f)+E_f V_f\right]^2} E_c \sigma_0 A \tag{13}$$

When $F' = 0$, then

$$(E_c - E_f)V_f^2 - 2E_c V_f + E_c = 0 \tag{14}$$

When $E_c = E_f$, the F is maximum, and $V_f = 0.5$; when $E_c \neq E_f$, for the equation

$$V_f^2 - \frac{2E_c}{E_c - E_f} V_f + \frac{E_c}{E_c - E_f} = 0 \tag{15}$$

Assuming $a = \frac{E_C}{E_C - E_f}$, since $E_c > 0$ and $E_f > 0$, then $a < 0$ or $a > 1$, thus $\Delta = 4a^2 - 4a > 0$ and Equation (15) has two different real roots, which are

$$V_f = a \pm \sqrt{a^2 - a} = \frac{E_c \pm \sqrt{E_c E_f}}{E_c - E_f} = \frac{1 \pm \sqrt{E_f/E_c}}{1 - E_f/E_c} \tag{16}$$

Since the real root $V_f = \frac{1+\sqrt{E_f/E_C}}{1-E_f/E_C}$ does not satisfy the condition $0 < V_f < 1$, it should be discarded, while the other real root

$$V_f = \frac{E_c - \sqrt{E_c E_f}}{E_c - E_f} \tag{17}$$

satisfies the condition and gives the maximum axial force storage F_{max}.

4.2. *Mechanical Model of CIF Composites*

The permanent anchor ends of CIF are the portion of the core fiber not coated with the corrodible coating, or the portion where the surface of the core fiber with corrodible coating is coated with the corrosion-resistant coating; the length of any permanent anchor end is defined as l'. When the CIF reserved with permanent anchor ends is added into the matrix composite, the pre-stress released to the matrix composite can be predicted when the shrinkage of CIF stops.

4.2.1. Basic Assumption

In order to simplify the calculation of interaction between the CIF and the brittle matrix composite, the following assumptions are made:

1. The CIF is unidirectionally and uniformly arranged in the matrix composite;
2. The influence of the Poisson's ratio on the magnitude of the axial stress is disregarded;
3. The permanent anchor ends are tightly bonded with the matrix composite without slippage;
4. The force influence of the corrosion product of the corrodible coating is disregarded.

4.2.2. Calculation of Internal Force

The symbols used in this section are listed in Table A2.

After the cross-section of the corrodible coating is completely lost, the pre-stress released to the brittle matrix composite by the shrinkage of the core fiber is maximum. As the corrosion product does not participate in the force, the core fiber and the brittle matrix composite establish a final tensile–compressive equilibrium. According to Equation (9), it can be known that the pre-stress released to the matrix composite by the shrinkage of the core fiber is

$$\sigma_m^p = -\frac{E_m \frac{V_{f1}}{1-V_{c1}} \sigma_f^p}{E_2} = -\frac{E_m V_{f1} \sigma_f^p}{E_f V_{f1} + E_m V_m} \tag{18}$$

At this point, the tensile stress in the core fiber is

$$\sigma_{f1}^p = \frac{E_m V_m \sigma_f^p}{E_f V_{f1} + E_m V_m} \tag{19}$$

where E_2 is the composite elastic modulus of the brittle matrix composite with the core fiber, and $E_2 = E_f V_{f1}/(1 - V_{c1}) + E_m V_m/(1 - V_{c1})$, $V_{f1} + V_{c1} = V_s$, and $V_s + V_m = 1$.

4.2.3. Anchor Length of CIF

In order that the permanent anchor ends are reliable without slipping, a sufficient length is required. It is assumed that the bonding anchoring force of the permanent anchor end is $T_a = \tau\pi d l'$, and the drawing force of the CIF is $T_t = \sigma_{f1}^p \pi d^2/4$. According to the force equilibrium $T_a = T_t$, the following is obtained

$$l' = \frac{d\sigma_{f1}^p}{4\tau} \tag{20}$$

where τ is the bonding stress between the CIF and the matrix composite at the interface, and when the composition and properties of the matrix composite and the CIF are known, τ can be determined; l' is the anchor length (the length of one end) of the CIF in the brittle matrix composite, and d is the diameter of the cross-section of the anchor end.

Formula (19) is substituted into Equation (20) to obtain

$$l' = \frac{d\sigma_{f1}^{p}}{4\tau} = \frac{dE_m V_m \sigma_f^p}{4\tau(E_f V_{f1} + E_m V_m)} \tag{21}$$

If the permanent anchor end is reliable for effectively transferring the pre-stress to the matrix composite without slipping, then

$$l' \geq \frac{dE_m V_m \sigma_f^p}{4\tau(E_f V_{f1} + E_m V_m)} \tag{22}$$

Thus, for the given material parameters of the CIF and the matrix composite, and the given volume fraction, the minimum length of the reliable anchor end of the CIF can be confirmed by calculation.

5. Discussion

Based on the concept of CIF, the self-repair method of the CIF composites has obvious advantages. First, the CIF composites working in a corrosive environment are capable of self-repairing without external help and independent of temperature, so compared to the self-repair techniques using SMA or an electric field, the use of CIF in concrete can reduce the costs. Second, the larger the pre-stress stored in the core fiber, or the higher corrosion degree the corrodible coating encounters, the larger the crack closure force that can be released to the concrete, meaning a wider crack can be repaired; thus, compared to the self-healing or self-repair techniques using crystalline admixtures, microcapsules, or bacteria, the use of CIF in concrete can close cracks with a wider width range without compromising the strength of the concrete. Third, before the corrodible coating is corroded thoroughly, when the concrete is again cracked, the corrodible coating can continue to be corroded until the cracks are closed or the cross-section of the corrodible coating is completely lost. Fourth, the use of CIF in concrete can act as effective reinforcement both before and after corrosion.

Based on the derived mechanical models, we can predict the self-repair capacity of the CIF composite. For example, we set the material of the corrodible coating of the CIF to be iron, the core fiber of the CIF to be a copper-plated steel fiber or a nylon fiber with a diameter of 0.2 mm (regardless of copper plating amount), and set the matrix composite to be concrete in a chloride environment. The material parameters of CIF and concrete are listed in Table 1. According to Section 4.1.3, if $E_c = E_f$, the maximum axial force stored in the core fiber (F_{max}) is obtained when the volume fraction of the core fiber in the CIF was 50%; if $E_c \neq E_f$, the F_{max} is obtained when the volume fraction of the core fiber is determined according to Equation (17). Setting the amount of CIFs in concrete (V_s) to be 4V%, and assuming that the CIF is unidirectionally and uniformly arranged in the concrete, then the pre-stress stored in the core fiber (σ_f^p) and the maximum pre-stress released to the concrete (σ_m^p) are calculated according to Equations (10) and (18), respectively. The recoverable strain of the core fiber (ε_f^p) can be calculated according to $\varepsilon_f^p = \sigma_f^p / E_f$. The results are shown in Table 1.

From the above sample calculation, it can be shown that, after the iron coating of the CIF is lost due to chloride corrosion, the recovery strain of the core fiber is 0.5% for the steel fiber and 12.7% for the nylon fiber. That is, if the length of the CIF is 20 mm, the recovery strain can be 2.5 mm for the CIF with the nylon core fiber, which means, if the CIF concrete member is free of external force, the theoretical crack closure width can be up to 2.5 mm. It can also be seen from Table 1 that the maximum pre-stress released to the concrete is 18.6 MPa for the CIF with the steel core fiber and 24.5 MPa for the CIF with the nylon core fiber. Figure 5 shows that when the amount of CIFs (V_s) and the initial tensile stress of the core fiber (σ_0) continuously increase, the maximum compressive stress applied to concrete also continuously increases. Thus, for the given composition and properties of the concrete and the CIF, the crack closing force provided by the CIF to concrete can be increased by increasing V_s and σ_0.

Table 1. Basic material parameters and calculation results of CIF and concrete.

No.	Composition	Material	Parameters			σ_f^p (MPa)	ε_f^p	σ_m^p (MPa)
			E (GPa)	V_{f1}	σ_0 (MPa)			
1	Corrodible coating	Iron	200	50% × 4%	-	-	-	-
	Core fiber	Steel	200	50% × 4%	2000	1000	0.5%	−18.6
	Matrix	Concrete	35	96%	-	-	-	-
2	Corrodible coating	Iron	200	14% × 4%	-	-	-	-
	Core fiber	Nylon	5.4 [77]	86% × 4%	800	687	12.7%	−24.5
	Matrix	Concrete	35	96%	-	-	-	-

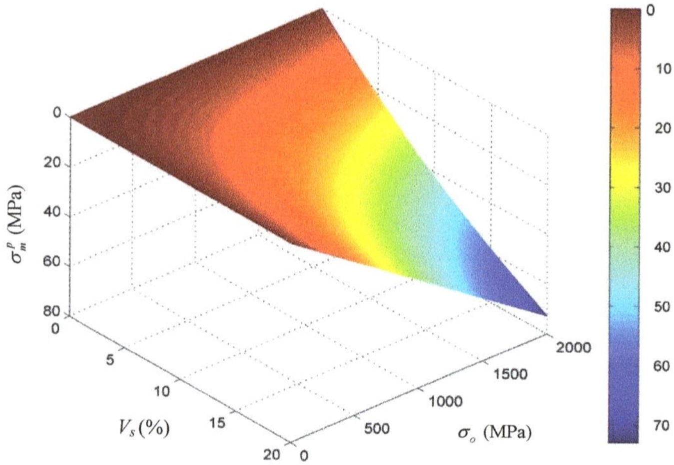

Figure 5. The influence of the amount of CIFs (V_s) and the initial tensile stress (σ_0) on the maximum pre-stress released to concrete (σ_m^p).

The application of CIF will benefit crack closure, increase rigidity, and improve the corrosion resistance of concrete. This paper only provides the conceptual and theoretical study of the CIF and CIF composites; many important issues remain for future study. First, the performance of the self-repairing system and the optimization of the composition and properties of the CIF need to be explained via finite element analysis. Second, the optimized CIF needs to be prepared in the laboratory and explored with a set of experiments that can show the self-repairing behavior of the CIF composite within a corrosive environment.

6. Conclusions

This paper presents a new concept of CIF that comprises an inner corrosion-resistant core fiber and an outer corrodible coating that can be easily corroded by corrosive media in the environment. During preparation process, the inner core fiber is put into tension and the outer corrodible coating into compression, such that the CIF is in equilibrium. When the CIF is in contact with corrosive media, the outer corrodible coating is corroded, and the core fiber shrinks and displays shape recovery, which in turn releases the pre-tension stress in the core fiber. By far, shape memory fibers comprising core fibers coated with a corrosion-resistant compound/material are well known, but a shape memory fiber by coating a corrodible coating on a core fiber has not yet been reported.

This paper also proposes a new self-repairing system that uses the CIFs to close cracks in brittle matrix composites within a corrosive environment. Once cracking occurs, the CIFs embedded in the matrix composite can be triggered to shrink by the corrosive media from the environment, which in turn releases the pre-stress stored in the core fiber and thereby applies a compressive force to the matrix composite that acts to close the cracks. By far, self-repair concrete comprising reinforcing fibers is well known, but not with a corrodible coating in equilibrium with the core fiber. Compared to the current self-repair or self-healing techniques for concrete, the use of CIF in concrete can cost less than using SMA or the electrochemical deposition method because it is independent of temperature and does not need external help. Furthermore, it can be more efficient for closing wider cracks than that provided by the crystalline admixtures, microcapsules, or bacteria methods, all of which have unfavorable effect on concrete strength. Additionally, the use of CIF in brittle matrix composite can act as effective reinforcement both before and after corrosion.

Based on the concepts, this paper also builds several mechanical models to predict the magnitude of pre-stress stored in the core fiber, the maximum pre-stress released to the brittle matrix composite, and the minimum length of the reliable anchor ends of CIF. These aim to attain an optimum combination of the CIF and matrix composite to provide enough crack closing force. Based on a sample calculation, the recovery strain was 0.5% for CIF with a steel core fiber and 12.7% for CIF with a nylon core fiber. The maximum crack closing force provided by the CIF to concrete can be increased by increasing the amount of CIFs in concrete and the initial tensile stress of the core fiber.

The presence of CIF can be helpful toward improving the crack resistance of concrete, especially the low-modulus polymer fiber concrete. It can help to reduce the probability of premature concrete cracking and improve the durability of the concrete structures in corrosive environments, including marine and underground environments. For the future work, many important issues related to the concepts need to be explored. First, the optimization of the composition and properties of the CIF needs to be found via finite element analysis before performing the time-consuming laboratory tests; second, the optimized CIF needs to be prepared in the laboratory; and third, a set of experiments should be conducted to explore the self-repairing behavior of the CIF composite within actual corrosive environments.

Author Contributions: Conceptualization, Z.W.; Methodology, Z.W.; Validation, Y.S.; Writing—original draft, Y.S. and D.W.; Writing—review and editing, Z.J. and J.S. All authors have read and agreed to the published version of the manuscript.

Funding: This research was funded by the National Natural Science Foundation of China, grant number 51808309; Mount Tai Shan Scholar of Shandong Province, grant number ts20190942; and the Key Technology Research and Development Program of Shandong Province, grant number 2019GGX102009.

Institutional Review Board Statement: Not applicable.

Data Availability Statement: No new data were created or analyzed in this study. Data sharing is not applicable to this article.

Conflicts of Interest: The authors declare no conflict of interest.

Appendix A.

Table A1. Symbols in formula deduction for the corrosion-induced intelligent fiber (CIF).

Symbol	Description
E_f	elastic modulus of the core fiber
E_c	elastic modulus of the corrodible coating
A_f	cross-sectional area of the core fiber
A_c	cross-sectional area of the corrodible coating
ε_f	initial tensile strain of the core fiber
ε_c	strain of the corrodible coating after equilibrium
V_f	volume fraction of the core fiber in the CIF
V_c	volume fraction of the corrodible coating in the CIF
E_1	elastic modulus of the CIF
σ_0	pre-tensile stress in the core fiber
σ_f^p	pre-stress stored in the core fiber after equilibrium
σ_c^p	compressive stress in the corrodible coating after equilibrium

Table A2. Symbols in formula deduction for CIF composites.

Symbol	Description
E_m	elastic modulus of the brittle matrix composite
E_2	composite elastic modulus of the matrix composite with the core fiber
V_{f1}	volume fraction of the core fiber in the CIF composite
V_{c1}	volume fraction of the corrodible coating in the CIF composite
V_s	volume fraction of the CIFs in the CIF composite
V_m	volume fraction of the matrix composite in the CIF composite
σ_m^p	pre-stress released to the brittle matrix composite after equilibrium
σ_{f1}^p	tensile stress in the core fiber after equilibrium

References

1. Jamari, J.; Ammarullah, M.I.; Santoso, G.; Sugiharto, S.; Supriyono, T.; van der Heide, E. In Silico Contact Pressure of Metal-on-Metal Total Hip Implant with Different Materials Subjected to Gait Loading. *Metals* **2022**, *12*, 1241. [CrossRef]
2. De Nardi, C.; Cecchi, A.; Ferrara, L. The Influence of Self-Healing Capacity of Lime Mortars on the Behaviour of Brick-Mortar Masonry Subassemblies. *Key Eng. Mater.* **2017**, *747*, 465–471. [CrossRef]
3. Maddalena, R.; Bonanno, L.; Balzano, B.; Tuinea-Bobe, C.; Sweeney, J.; Mihai, I. A crack closure system for cementitious composite materials using knotted shape memory polymer (k-SMP) fibres. *Cem. Concr. Compos.* **2020**, *114*, 103757. [CrossRef]
4. Greil, P. Self-Healing Engineering Ceramics with Oxidation-Induced Crack Repair. *Adv. Eng. Mater.* **2020**, *22*, 1901121. [CrossRef]
5. Albuhairi, D.; Di Sarno, L. Low-Carbon Self-Healing Concrete: State-of-the-Art, Challenges and Opportunities. *Buildings* **2022**, *12*, 1196. [CrossRef]
6. Hearn, N. *Saturated Permeability of Concrete as Influenced by Cracking and Self-Sealing*; University of Cambridge: Cambridge, UK, 1993.
7. Hearn, N.; Morley, C.T. Self-sealing property of concrete—Experimental evidence. *Mater. Struct./Mater. Constr.* **1997**, *30*, 404–411. [CrossRef]
8. Roig-Flores, M.; Serna, P. Concrete Early-Age Crack Closing by Autogenous Healing. *Sustainability* **2020**, *12*, 4476. [CrossRef]
9. Qureshi, T.; Al-Tabbaa, A. *Self-Healing Concrete and Cementitious Materials*; Chapter 12; IntechOpen: London, UK, 2020; pp. 191–214. [CrossRef]
10. De Belie, N.; Gruyaert, E.; Al-Tabbaa, A.; Antonaci, P.; Baera, C.; Bajare, D.; Darquennes, A.; Davies, R.; Ferrara, L.; Jefferson, T.; et al. A Review of Self-Healing Concrete for Damage Management of Structures. *Adv. Mater. Interfaces* **2018**, *5*, 1800074. [CrossRef]
11. Edvardsen, C. Water Permeability and Autogenous Healing of Cracks in Concrete. *Mater. J.* **1999**, *96*, 448–454. [CrossRef]
12. Ziegler, F.; Masuero, A.B.; Pagnussat, D.T.; Molin, D.C.C.D. Evaluation of Internal and Superficial Self-Healing of Cracks in Concrete with Crystalline Admixtures. *Materials* **2020**, *13*, 4947. [CrossRef]
13. Li, G.; Huang, X.; Lin, J.; Jiang, X.; Zhang, X. Activated chemicals of cementitious capillary crystalline waterproofing materials and their self-healing behaviour. *Constr. Build. Mater.* **2019**, *200*, 36–45. [CrossRef]
14. Ravitheja, A.; Reddy, T.; Sashidhar, C. Self-healing concrete with crystalline admixture—A review. *J. Wuhan Univ. Technol.-Mater. Sci. Ed.* **2019**, *34*, 1143–1154. [CrossRef]
15. Zhang, C.; Lu, R.; Li, Y.; Guan, X. Effect of crystalline admixtures on mechanical, self-healing and transport properties of engineered cementitious composite. *Cem. Concr. Compos.* **2021**, *124*, 104256. [CrossRef]

16. Xue, C.; Li, W.; Qu, F.; Sun, Z.; Shah, S.P. Self-healing efficiency and crack closure of smart cementitious composite with crystalline admixture and structural polyurethane. *Constr. Build. Mater.* **2020**, *260*, 119955. [CrossRef]
17. Žáková, H.; Pazderka, J.; Reiterman, P. Textile reinforced concrete in combination with improved self-healing ability caused by crystalline admixture. *Materials* **2020**, *13*, 5787. [CrossRef]
18. Sisomphon, K.; Copuroglu, O.; Koenders, E. Self-healing of surface cracks in mortars with expansive additive and crystalline additive. *Cem. Concr. Compos.* **2012**, *34*, 566–574. [CrossRef]
19. De Souza Oliveira, A.; Gomes, O.d.F.M.; Ferrara, L.; Fairbairn, E.d.M.R.; Toledo Filho, R.D. An overview of a twofold effect of crystalline admixtures in cement-based materials: From permeability-reducers to self-healing stimulators. *J. Build. Eng.* **2021**, *41*, 102400. [CrossRef]
20. Roig-Flores, M.; Pirritano, F.; Serna, P.; Ferrara, L. Effect of crystalline admixtures on the self-healing capability of early-age concrete studied by means of permeability and crack closing tests. *Constr. Build. Mater.* **2016**, *114*, 447–457. [CrossRef]
21. Song, G.; Ma, N.; Li, H.N. Applications of shape memory alloys in civil structures. *Eng. Struct.* **2006**, *28*, 1266–1274. [CrossRef]
22. Siddiquee, K.N.; Billah, A.H.M.M.; Issa, A. Seismic collapse safety and response modification factor of concrete frame buildings reinforced with superelastic shape memory alloy (SMA) rebar. *J. Build. Eng.* **2021**, *42*, 102468. [CrossRef]
23. Abdulridha, A.; Palermo, D.; Foo, S.; Vecchio, F.J. Behavior and modeling of superelastic shape memory alloy reinforced concrete beams. *Eng. Struct.* **2013**, *49*, 893–904. [CrossRef]
24. Kuang, Y.; Ou, J. Self-repairing performance of concrete beams strengthened using superelastic SMA wires in combination with adhesives released from hollow fibers. *Smart Mater. Struct.* **2008**, *17*, 025020. [CrossRef]
25. Mohd Jani, J.; Leary, M.; Subic, A.; Gibson, M.A. A review of shape memory alloy research, applications and opportunities. *Mater. Des.* **2014**, *56*, 1078–1113. [CrossRef]
26. Furuya, Y.; Sasaki, A.; Taya, M. Enhanced mechanical properties of TiNi shape memory fiber/Al matrix composite. *Mater. Trans. JIM* **1993**, *34*, 224–227. [CrossRef]
27. Chen, W.; Feng, K.; Wang, Y.; Lin, Y.; Qian, H. Evaluation of self-healing performance of a smart composite material (SMA-ECC). *Constr. Build. Mater.* **2021**, *290*, 123216. [CrossRef]
28. Cohades, A.; Hostettler, N.; Pauchard, M.; Plummer, C.J.G.; Michaud, V. Stitched shape memory alloy wires enhance damage recovery in self-healing fibre-reinforced polymer composites. *Compos. Sci. Technol.* **2018**, *161*, 22–31. [CrossRef]
29. Geetha, S.; Selvakumar, M. A composite for the Future-Concrete Composite Reinforced with Shape Memory Alloy Fibres. *Mater. Today Proc.* **2019**, *18*, 5550–5555. [CrossRef]
30. Jia, Y.-q.; Lu, Z.-D.; Li, L.-Z.; Wu, Z.-L. A Review of Applications and Research of Shape Memory Alloys in Civil Engineering. *IOP Conf. Ser. Mater. Sci. Eng.* **2018**, *392*, 022009. [CrossRef]
31. Dry, C. Procedures developed for self-repair of polymer matrix composite materials. *Compos. Struct.* **1996**, *35*, 263–269. [CrossRef]
32. White, S.R.; Sottos, N.R.; Geubelle, P.H.; Moore, J.S.; Kessler, M.R.; Sriram, S.; Brown, E.N.; Viswanathan, S. Autonomic healing of polymer composites. *Nature* **2001**, *409*, 794–797. [CrossRef]
33. Zotiadis, C.; Patrikalos, I.; Loukaidou, V.; Korres, D.M.; Karantonis, A.; Vouyiouka, S. Self-healing coatings based on poly(urea-formaldehyde) microcapsules: In situ polymerization, capsule properties and application. *Prog. Org. Coat.* **2021**, *161*, 106475. [CrossRef]
34. Rule, J.D.; Sottos, N.R.; White, S.R. Effect of microcapsule size on the performance of self-healing polymers. *Polymer* **2007**, *48*, 3520–3529. [CrossRef]
35. Wang, J.Y.; Soens, H.; Verstraete, W.; De Belie, N. Self-healing concrete by use of microencapsulated bacterial spores. *Cem. Concr. Res.* **2014**, *56*, 139–152. [CrossRef]
36. Dubey, R.; Shami, T.C.; Bhasker Rao, K.U. Microencapsulation technology and applications. *Def. Sci. J.* **2009**, *59*, 82–95. [CrossRef]
37. Bonilla, L.; Hassan, M.M.; Noorvand, H.; Rupnow, T.; Okeil, A. Dual self-healing mechanisms with microcapsules and shape memory alloys in reinforced concrete. *J. Mater. Civ. Eng.* **2018**, *30*, 04017227. [CrossRef]
38. Aghamirzadeh, G.R.; Khalili, S.M.R.; Eslami-Farsani, R.; Saeedi, A. Experimental investigation on the smart self-healing composites based on the short hollow glass fibers and shape memory alloy strips. *Polym. Compos.* **2019**, *40*, 1883–1889. [CrossRef]
39. Li, W.; Jiang, Z.; Yu, Q. Multiple damaging and self-healing properties of cement paste incorporating microcapsules. *Constr. Build. Mater.* **2020**, *255*, 119302. [CrossRef]
40. Wang, Y.; Lin, Z.; Tang, C.; Hao, W. Influencing factors on the healing performance of microcapsule self-healing concrete. *Materials* **2021**, *14*, 4139. [CrossRef]
41. Jiang, S.; Lin, Z.; Tang, C.; Hao, W. Preparation and mechanical properties of microcapsule-based self-healing cementitious composites. *Materials* **2021**, *14*, 4866. [CrossRef]
42. Castaneda, H.; Hassan, M.M.; Radovic, M.; Milla, J.; Karayan, A. *Self-Healing Microcapsules as Concrete Aggregates for Corrosion Inhibition in Reinforced Concrete*; Transportation Consortium of South-Central States: Washington, DC, USA, 2018.
43. Pulikkalparambil, H.; Sanjay, M.; Siengchin, S.; Khan, A.; Jawaid, M.; Pruncu, C.I. Self-repairing hollow-fiber polymer composites. In *Woodhead Publishing Series in Composites Science and Engineering, Self-Healing Composite Materials*; Woodhead Publishing: Sawston, UK, 2020; pp. 313–326. [CrossRef]
44. Feng, J.; Dong, H.; Wang, R.; Su, Y. A novel capsule by poly (ethylene glycol) granulation for self-healing concrete. *Cem. Concr. Res.* **2020**, *133*, 106053. [CrossRef]

45. Luo, M.; Qian, C.-X.; Li, R.-Y. Factors affecting crack repairing capacity of bacteria-based self-healing concrete. *Constr. Build. Mater.* **2015**, *87*, 1–7. [CrossRef]
46. Luo, M.; Qian, C. Influences of bacteria-based self-healing agents on cementitious materials hydration kinetics and compressive strength. *Constr. Build. Mater.* **2016**, *121*, 659–663. [CrossRef]
47. Van Tittelboom, K.; De Belie, N.; De Muynck, W.; Verstraete, W. Use of bacteria to repair cracks in concrete. *Cem. Concr. Res.* **2010**, *40*, 157–166. [CrossRef]
48. Wiktor, V.; Jonkers, H.M. Quantification of crack-healing in novel bacteria-based self-healing concrete. *Cem. Concr. Compos.* **2011**, *33*, 763–770. [CrossRef]
49. Intarasoontron, J.; Pungrasmi, W.; Nuaklong, P.; Jongvivatsakul, P.; Likitlersuang, S. Comparing performances of MICP bacterial vegetative cell and microencapsulated bacterial spore methods on concrete crack healing. *Constr. Build. Mater.* **2021**, *302*, 124227. [CrossRef]
50. Allahyari, H.; Heidarpour, A.; Shayan, A. Experimental and analytical studies of bacterial self-healing concrete subjected to alkali-silica-reaction. *Constr. Build. Mater.* **2021**, *310*, 125149. [CrossRef]
51. Zhang, X.; Qian, C. Engineering application of microbial self-healing concrete in lock channel wall. *Mar. Georesources Geotechnol.* **2022**, *40*, 96–103. [CrossRef]
52. Sri Durga, C.S.; Ruben, N.; Sri Rama Chand, M.; Indira, M.; Venkatesh, C. Comprehensive microbiological studies on screening bacteria for self-healing concrete. *Materialia* **2021**, *15*, 101051. [CrossRef]
53. Chen, X.; Yuan, J.; Alazhari, M. Effect of microbiological growth components for bacteria-based self-healing on the properties of cement mortar. *Materials* **2019**, *12*, 1303. [CrossRef]
54. Van Wylick, A.; Monclaro, A.V.; Elsacker, E.; Vandelook, S.; Rahier, H.; De Laet, L.; Cannella, D.; Peeters, E. A review on the potential of filamentous fungi for microbial self-healing of concrete. *Fungal Biol. Biotechnol.* **2021**, *8*, 16. [CrossRef]
55. Feng, C.; Zong, X.; Cui, B.; Guo, H.; Zhang, W.; Zhu, J. Application of Carrier Materials in Self-Healing Cement-Based Materials Based on Microbial-Induced Mineralization. *Crystals* **2022**, *12*, 797. [CrossRef]
56. Onyelowe, K.C.; Ebid, A.M.; Riofrio, A.; Baykara, H.; Soleymani, A.; Mahdi, H.A.; Jahangir, H.; Ibe, K. Multi-Objective Prediction of the Mechanical Properties and Environmental Impact Appraisals of Self-Healing Concrete for Sustainable Structures. *Sustainability* **2022**, *14*, 9573. [CrossRef]
57. Eran, Y.C.; Hernandez-Sanabria, E.; Boon, N.; De Belie, N. Enhanced crack closure performance of microbial mortar through nitrate reduction. *Cem. Concr. Compos.* **2016**, *70*, 159–170. [CrossRef]
58. Palin, D.; Wiktor, V.; Jonkers, H.M. A bacteria-based self-healing cementitious composite for application in low-temperature marine environments. *Biomimetics* **2017**, *2*, 13. [CrossRef]
59. Erşan, Y.Ç.; Gruyaert, E.; Louis, G.; Lors, C.; De Belie, N.; Boon, N. Self-protected nitrate reducing culture for intrinsic repair of concrete cracks. *Front. Microbiol.* **2015**, *6*, 1228. [CrossRef]
60. Chu, H.-Q.; Jiang, L.-H.; Yu, L. Effect of electrodeposition on BET surface area and microstructure of cement mortar. *J. Build. Mater.* **2006**, *9*, 627–632. (In Chinese)
61. Goux, A.; Pauporte, T.; Chivot, J.; Lincot, D. Temperature effects on ZnO electrodeposition. *Electrochim. Acta* **2005**, *50*, 2239–2248. [CrossRef]
62. Otsuki, N.; Ryu, J.S. Use of electrodeposition for repair of concrete with shrinkage cracks. *J. Mater. Civ. Eng.* **2001**, *13*, 136–142. [CrossRef]
63. Kim, J.-K.; Yee, J.-J.; Kee, S.-H. Electrochemical deposition treatment (Edt) as a comprehensive rehabilitation method for corrosion-induced deterioration in concrete with various severity levels. *Sensors* **2021**, *21*, 6287. [CrossRef]
64. Ryu, J.-S.; Otsuki, N. Crack closure of reinforced concrete by electrodeposition technique. *Cem. Concr. Res.* **2002**, *32*, 159–164. [CrossRef]
65. Yan, Z.; Chen, Q.; Zhu, H.; Woody Ju, J.; Zhou, S.; Jiang, Z. A multi-phase micromechanical model for unsaturated concrete repaired using the electrochemical deposition method. *Int. J. Solids Struct.* **2013**, *50*, 3875–3885. [CrossRef]
66. Liu, L.; Dong, B. Investigation of concrete crack repair by electrochemical deposition. *Int. J. Electrochem. Sci.* **2021**, *16*, 2. [CrossRef]
67. Wang, Y.; Wang, C.; Zhou, S.; Liu, K. Influence of Anode Material on the Effect of Electrophoretic Deposition for the Repair of Rust-Cracked Reinforced Concrete. *Constr. Build. Mater.* **2022**, *335*, 127466. [CrossRef]
68. Zhu, H.; Chen, Q.; Ju, J.W.; Yan, Z.; Jiang, Z. Electrochemical deposition induced continuum damage-healing framework for the cementitious composite. *Int. J. Damage Mech.* **2021**, *30*, 945–963. [CrossRef]
69. Yang, Q.; Wang, J.; Yuan, L.; Zhou, Z. Effect of graphene and carbon fiber on repairing crack of concrete by electrodeposition. *Ceram.-Silik.* **2019**, *63*, 403–412. [CrossRef]
70. Wang, C.; Wang, Y.; Liu, K.; Zhou, S. Effect of colloid solution concentration of epoxy resin on properties of rust-cracked reinforced concrete repaired by electrophoretic deposition. *Constr. Build. Mater.* **2022**, *318*, 126184. [CrossRef]
71. Cuenca, E.; Ferrara, L. Self-healing capacity of fiber reinforced cementitious composites. State of the art and perspectives. *KSCE J. Civ. Eng.* **2017**, *21*, 2777–2789. [CrossRef]
72. Liu, H.; Zhang, Q.; Gu, C.; Su, H.; Li, V. Influence of microcrack self-healing behavior on the permeability of Engineered Cementitious Composites. *Cem. Concr. Compos.* **2017**, *82*, 14–22. [CrossRef]
73. Wang, D.; Ju, Y.; Shen, H.; Xu, L. Mechanical properties of high performance concrete reinforced with basalt fiber and polypropylene fiber. *Constr. Build. Mater.* **2019**, *197*, 464–473. [CrossRef]

74. Balzano, B.; Sweeney, J.; Thompson, G.; Tuinea-Bobe, C.-L.; Jefferson, A. Enhanced concrete crack closure with hybrid shape memory polymer tendons. *Eng. Struct.* **2021**, *226*, 111330. [CrossRef]
75. Garg, M.; Azarsa, P.; Gupta, R. Self-healing potential and post-cracking tensile behavior of polypropylene fiber-reinforced cementitious composites. *J. Compos. Sci.* **2021**, *5*, 122. [CrossRef]
76. Wei, M.; Zhong, K.; Cai, S.; Feng, T.; Zhang, N. Preparation and application of the novel resin microcapsules in self-repairing cement composites with polypropylene fibers. *J. Adhes. Sci. Technol.* **2022**, 1–20. [CrossRef]
77. Ahmad, J.; Zaid, O.; Aslam, F.; Shahzaib, M.; Ullah, R.; Alabduljabbar, H.; Khedher, K.M. A study on the mechanical characteristics of glass and nylon fiber reinforced peach shell lightweight concrete. *Materials* **2021**, *14*, 4488. [CrossRef] [PubMed]

polymers

MDPI

Review

Durability of CFRP–Steel Double–Lap Joints under Cyclic Freeze–Thaw/Wet–Dry Conditions

Xiang Ren [1,*], Lingzhi Jiang [1], Jun He [2], Yi Yang [1,†], Yamin Sun [1], Qunfeng Liu [1] and Shaojie Chen [1]

1 School of Architecture and Civil Engineering, Xi'an University of Science and Technology, Xi'an 710054, China
2 School of Civil Engineering, Chang'an University, Xi'an 710061, China
* Correspondence: renxiang798@xust.edu.cn
† Current institution: China Construction Xi'an Investment & Development Co., Ltd., Xi'an 710000, China.

Abstract: The usage of carbon fiber–reinforced polymer (CFRP) to strengthen cracked steel structures can effectively improve its bear capacity, so it has been extensively used in recent years. The degradation of interfacial bonding is one of the most important factors affecting the durability of CFRP–steel structures under a freeze–thaw(F–T)/wet–dry (W–D) environment. In this study, epoxy resin adhesive (ERA) dog-bone specimens and CFRP–steel double-lap joints (bonded joints) were prepared. F–T/W–D cycles experiment and tensile tests of the ERA specimens and the bonded joints were also performed. Under F–T/W–D cycles, the main properties of the ERA specimens and the bonded joints were examined. Results indicated that fracture failure occurred in all ERA specimens. The hybrid failure modes of fiber peeling on the surface of CFRP plate and the bonded interface peeling between the CFRP plate and ERA layer primarily occurred in the bonded joints. The failure of both of them can be considered to be brittle, which was unaffected by the F–T/W–D cycles. With increased F–T/W–D cycles, the ultimate load and tensile strength of the ERA specimens initially increased and then decreased, whereas the elastic modulus initially increased and then remained unchanged. The ultimate load of the bonded joints decreased gradually. Based on the relationship between the interfacial bond-slip parameters and the number of F–T/W–D cycles, the bond–slip model of the bonded joints was established. The proposed relationship was validated by comparing with the experimental bond-slip relationships and the predicted relationships under the F–T/W–D cycles.

Keywords: CFRP–steel joints; freeze–thaw/wet–dry environment; failure mode; ultimate load; degradation model

1. Introduction

With the continuous increase in the total weight of operating vehicles and highway-traffic flow, the fatigue–cracking problem of existing steel bridges is becoming increasingly serious [1,2]. Drilling holes, welding, and additive steel-plate reinforcement methods are the traditional reinforcement or repair methods for cracked steel bridges. However, these methods have inherent defects, such as increased welding residual stress, local stress, and additional weight [3]. Carbon fiber-reinforced polymer (CFRP) has excellent properties such as light weight, high tensile strength, fatigue resistance, corrosion resistance, and convenience for construction. When it is attached onto the cracked part of the steel bridge, it can effectively improve the mechanical performance and does not excessively increase the dead weight of the bridge. Accordingly, the externally bonded CFRP–reinforcement technology has been used to rehabilitate steel bridges in recent years and has achieved good results [4–7]. The durability of rehabilitated structures by bonding CFRP plates depends on the bond performance of the reinforcement interface rather than the strength of the CFRP material itself. When the reinforced structures are exposed to the action of wet–dry(W–D), freeze-thaw(F–T) and other harsh environments for a long time, the performance of the bonded interface is reduced [8–12].

Epoxy resin adhesive (ERA) material is the main component of the bonded interface of CFRP–strengthened bridge structures, and its performance directly affects the bond properties of CFRP–steel double-lap joints [13,14]. The ERA is sensitive to ambient temperature and humidity [15], and it typically suffers from brittle failure, especially in cold environments [16,17]. Tensile strength also degrades with increased F–T cycle [18,19]. After reaching a certain cycle number, the tensile–shear strength returns to the unfrozen state [20]. The change in Young's modulus is insignificant [21], and the elastic modulus of the ERA is hardly affected after 250 F–T cycles (−17 °C–8 °C) [22]. However, relevant studies have demonstrated that the tensile strength and elastic modulus of ERA increases after the first 50 F–T cycles owing to post-curing effect and continuously decreases after the second 50 F–T cycles [23]. The tensile-strength promotion of the ERA is primarily due to the release of internal stress caused by the diffusion of water molecules inside it [24].

Similarly, the performance of ERA also degrades when exposed to chlorine solution, seawater, and brine at different temperatures. The tensile strength and stiffness of the ERA specimens degrade at a high rate under seawater conditions of 20 °C and 50 °C [25], as well as brine at 40 °C for a long time [26]. Compared with fiber materials, the tensile properties of epoxy resin matrix are more sensitive to W–D exposure [27]. However, some studies have concluded that the influence on the performance of the ERA is minimal when exposed to chlorine solution for 36 months [28]. The salt solution accelerates the degradation of the ERA, and the presence of chlorine salt in F–T exposure aggravates the degradation [29].

CFRP–steel double-lap joints or single-lap joint specimens are often used to simulate steel bridges strengthened with CFRP. Borrie et al. [11] submerged CFRP–steel double-lap joints in a 5% NaCl solution for different durations up to 6 months at 20 °C, 40 °C and 50 °C, and subsequently conducted static tensile tests. Results show that overall, more than half of the specimens' strength decreases by at least a 10%. In seawater at 20 °C and 50 °C, the degradation rate of strength and stiffness of the CFRP–steel specimens primarily occur in the initial stage, consistent with the degradation law of the ERA. This finding indicates that the degradation of the CFRP–steel specimens is primarily affected by the ERA [25]. In distilled water (20 °C, 45 °C) and 5% sodium chloride solution, the tensile strength and elastic modulus of the ERA decrease linearly with increased water content, but the ultimate load of FRP–steel specimen increase after soaking in distilled water and salt solution at 20 °C for one year [30]. In W–D cycle environment, the bear capacity of CFRP–steel beam decreases significantly and tends to decrease continually with increased W–D cycles [9]. Some studies have shown that the bearing capacity of CFRP–steel specimen tends to be stable after reaching a certain number of W–D cycles [31]. Bond-strength degradation is likely owing to the degradation of the bond property between the steel and the ERA layer. ERA-property degradation has also contributed to the property degradation of CFRP–steel specimens, but the role of this mechanism is apparently secondary [32,33].

Under F–T cycle condition (−16 °C–38 °C, 100% RH), the strength-reduction rate of CFRP–steel single-lap joints specimens in the early stage of F–T cycle is obviously higher than that in the late stage. The failure–mode change with increased F–T cycles [8]. The number of F–T cycles has no effect on the failure mode, the curve shapes of the interfacial shear stress distribution, and the bond–slip relationship of the steel-FRP lap joints. The bond–slip relationship curve is also approximately bilinear. However, the mechanical properties of the ERA and the steel–FRP lap joints specimens decrease with increased F–T cycles [34]. Conversely, other researchers have revealed that the tensile strength and elastic modulus of an ERA, Araldite–2015 (CN), and the ultimate load of the CFRP–steel specimen initially increase slightly and then decrease with increased F–T cycles (−20 °C–8 °C). The failure mode of all CFRP–steel specimens is mixed failure [23]. Yang et al. [35] studied the aging test of an ERA, SIKA30, and CFRP–steel joint in F–T environment (−20 °C–20 °C, 50% RH). Results showed that the tensile strength, elastic modulus, and ultimate strain of the ERA and the interfacial bond strength of the CFRP–steel specimen do not decrease after 10,000 h (883 cycles). However, the failure mode of the CFRP–steel specimen changes from cohesive failure to mixed failure after 10,000 h.

Heshmati et al. [36] soaked ERA and CFRP–steel specimens in 45 °C water or 45 °C saltwater for 36 months and then performed F–T experiments. Results show that the ERA ductility decreases after being dried from a wet state. However, such reduction is restored after the second absorption cycle. The failure mode of the CFRP–steel specimen pretreated with the salt solution changes, and the failure loads do not change significantly. This finding is due to the influence of water and temperature. The strength of CFRP–steel joints decreases after a complete W–D cycle in distilled water and saltwater, respectively. Their strengths are significantly larger than those observed after only wet exposure. Moreover, 125 and 250 F–T cycles are found to have no unfavorable effect on the strength of dry or preconditioned joints. Kim et al. [37] studied the bear capacity of FRP–steel joints under W–D and F–T cycles and found that the bond strength of components increases by 31.8% and 17.2%, respectively, after 100 cycles of W–D and F–T owing to the post-curing effect of water [33]. The interfacial stiffness of the specimen gradually decreases under the effects of W–D and F–T. The degradation of the specimen under the F–T cycle is more significant than that of W–D cycle. The failure mode of the specimen changes from cohesive failure to debonding failure.

In summary, ERA degradation in W–D and F–T environment is generally believed to be primarily caused by the hydrolysis of the ERA under the action of water molecules and the reversible softening reaction, thereby affecting the properties of the ERA [22,32,35]. For the bonded joints, first, the degradation of the bond strength and stiffness are controlled by bond-degradation properties of the bonded interface, although the deterioration rate of the strength and stiffness of the ERA and the bonded joints are basically the same. Second, the F–T, W–D cycles environment exert a dominant effect on the bond-property degradation of the bonded joints, and the effect is obviously greater on the ERA properties.

China has a vast territory and changeable climate. The south is wet and rainy, whereas the north is cold and dry. The cold region is extensively distributed, and the seasonal F–T environment is particularly obvious. Freeze and thaw environment in cold regions exist obviously with manmade chloride-corrosion environment by deicing method of salting snow on bridge deck in winter [38]. The action of salt corrosion and F–T environment leads to the infiltration of liquid into the interface of the bonded joints. The repeated action of salt corrosion, frost heaving, and melting aggravates the extension and propagation of microcracks in the bonded interface, which will seriously affect the service life of the bridge. Considering that the actual reinforced bridges continue to be affected by the effects of F–T/W–D, it is very necessary to study the long-term performance of CFRP steel double lap joints in the F–T/W–D environment.

The present study investigated the bond properties of the ERA and the bonded joints under the F–T/W–D interaction cycles through tests of the ERA and bond joints. The failure mode, ultimate load, and surface strain of the ERA and the bonded joints were tested. The main objectives of this study were as follow: (1) to study the effect of the F–T/W–D interaction cycle on the mechanical properties of the ERA and the bonded joints; (2) to analyze bond–slip relationship of the bonded joints under F–T/W–D interaction cycle; and (3) to simulate bond–slip degradation relationship of the bonded joints considering the F–T/W–D interaction cycle. The results can provide the basis for the development of the predictive degradation of ERA materials and the design of CFRP–strengthened structures.

2. Experimental Program

2.1. Material Properties

The steel plates used in the bonded joints were Q345B, whose characteristics were obtained by referring to the Chinese standard [39]. The ERA material was a mixture of two components A and B with a mass ratio of 2:1. The CFRP plate and ERA were produced by Nanjing Mankat Technology Co., Ltd., Nanjiang, Jiangsu Province, China, and their mechanical properties were provided by the manufacturer. Table 1 shows the material properties.

Table 1. The mechanical properties of the materials.

Material Type	Tensile Strength (MPa)	Tensile Modulus (GPa)	Elongation Break (%)	Shear Strength (MPa)
CFRP	2400	160	1.65	50.5
ERA	30.3	4.92	1.62	/
Steel	455	206	/	/

2.2. Specimens and Test Procedures

2.2.1. ERA

A total of 15 specimens were prepared to investigate the effect of the F–T/W–D cycles on the mechanical properties of the ERA, as shown in Figure 1a. The ERA specimens were made according to GB/T2567–2008 [40]. According to the manufacturer's recommendations, the ERA specimens were cured for at least 7 d at room temperature before testing. Three control specimens were tested directly after the ERA specimens were fully cured. Other ERA specimens were subjected to the F–T/W–D cycles after they were cured and subsequently tested according to ASTM D638–10 [41]. The ERA specimen identifications could be used the type of *Xi-m*, where X represents the F–T/W–D action, *i* is the number of F–T/W–D cycles, *m* is the number of specimen.

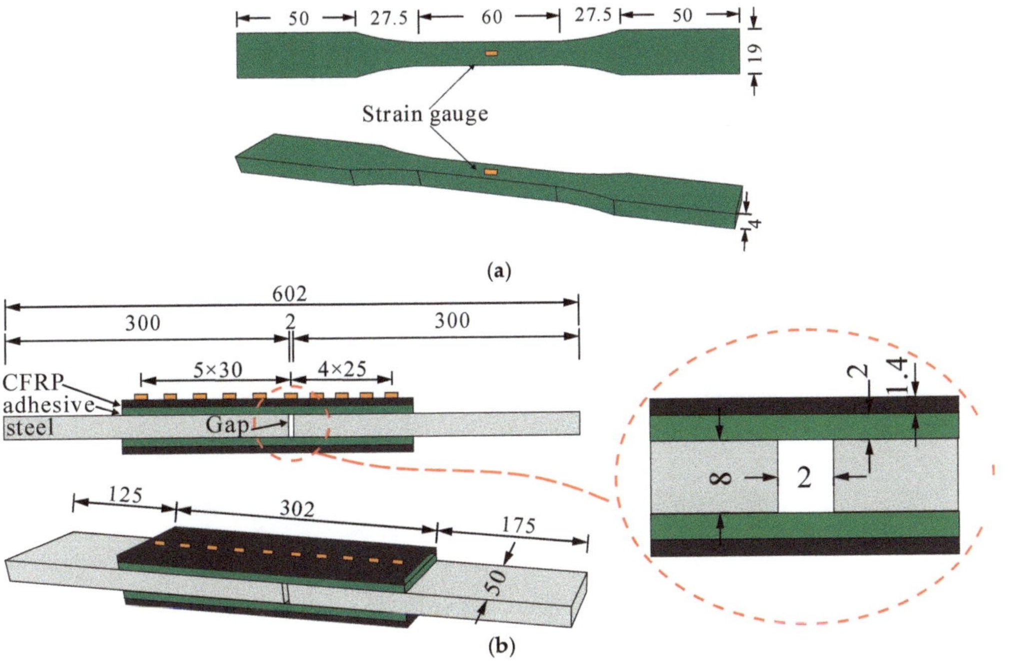

Figure 1. Geometry and strain gauge arrangement of the specimens. (Unit: mm). (**a**) The ERA specimen. (**b**) The CFRP–steel specimen.

2.2.2. Bonded Joints

A total of 15 specimens were made to study the effect of the F–T/W–D cycles on the bond properties of the bonded joints. The bonded-joints specimens were prepared according to ASTM D3528–96 (2008) and were tested according to ASTM D638–10 [40]. For the bonded joints, the length, width, and thickness of steel plates were 300, 50, and 8 mm, respectively. CFRP plates had identical length and width to those of steel plates. The thickness of the CFRP plates and ERA were 1.4 and 2.0 mm, respectively. The joint interval of steel plates was 2 mm, and the CFRP plate adopted asymmetric-bonding method. The bonding lengths on each side of the joints were 175 and 125 mm. The initial peeling of the bond interface was expected to occur in the shorter side direction, as shown in Figure 1b. The designation of the CFRP–steel joints was the same as that of the ERA specimens.

2.2.3. Specimen Preparation

The interface-treatment method is one of the most important factors affecting the bond properties of bonded joints [42]. To improve the bond performance of the bonded joints, the steel plate surface was sandblasted [43], and the CFRP plate surface was evenly polished with 100 mesh fine sandpaper along the distribution direction of CFRP fiber before preparing the specimens. Then, the steel plate and CFRP plate surface were cleaned with acetone to ensure dryness and absence of dust. Afterwards, the CFRP plate was pasted on the both sides of the steel plate with 2 mm interval, and cured at room temperature in a specific mold for 7 d.

2.2.4. Environmental Condition

In the F–T/W–D cycle experiment, F–T cycle was first conducted followed by W–D cycle. The environmental cycles were 0, 30, 60, 90, and 120 times. Cycle 0 represented no F–T/W–D cycle experienced and was defined as the control group. According to the slow F–T method in the Chinese National Standard GB/T-50082 [44], the temperature of the F–T cycle ranged from −20 °C to +30 °C, and the change from +30 °C to −20 °C (−20 °C to +30 °C) took 30 min at a rate of 1.2 °C/min. The time of the F–T cycle was around 8 h with 4 h of freezing and 4 h of thawing. The W–D environment was applied for 8 h per cycle while being soaked in 5% NaCl solution at room temperature for 4 h and air drying at room temperature for 4 h.

2.2.5. Test Setup and Procedure

A WAW-300 microcomputer servo tester by Shenzhen Wance testing machine equipment Co., Ltd., Shenzhen, Guangzhou province, China, was adopted to conduct static tensile tests on ERA specimens and the bonded joints. Before loading, the upper and lower beams were adjusted, and the tensile force area of the specimen was completely exposed. Then, the upper and lower clips were adjusted to the clamp area of the specimen. The tensile–loading device is shown in Figure 2. The tensile test was loaded in a displacement-controlled manner to stretch the specimen at a tensile rate of 0.5 mm/min. To prevent the specimen from sliding between the upper and lower clips, the preload for 0.1–0.2 kN was performed. The failure mode, ultimate load, and surface strain of CFRP were recorded during tensile test.

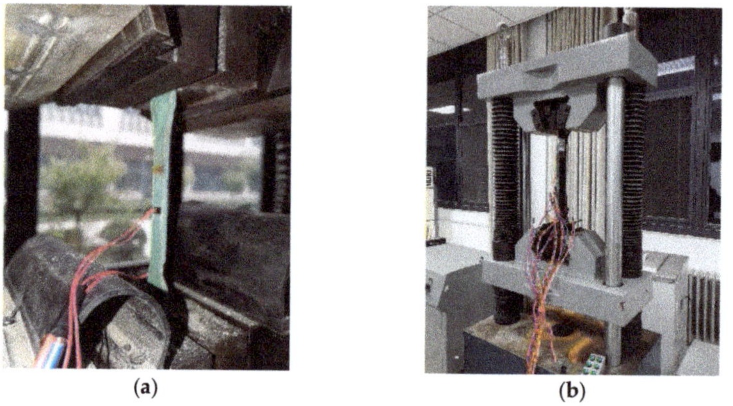

(a) (b)

Figure 2. Schematic diagram of test setup. (**a**) The ERA specimen. (**b**) The bonded joints.

3. Experimental Results

3.1. ERA Specimen

3.1.1. Failure Modes

The typical failure modes of the ERA specimen are shown in Figure 3. As shown in Figure 3, no obvious difference existed in the failure modes of all ERA specimens. The specimens had no neck shrinkage in the tensile process, and the fracture was sudden, indicating brittle characteristics.

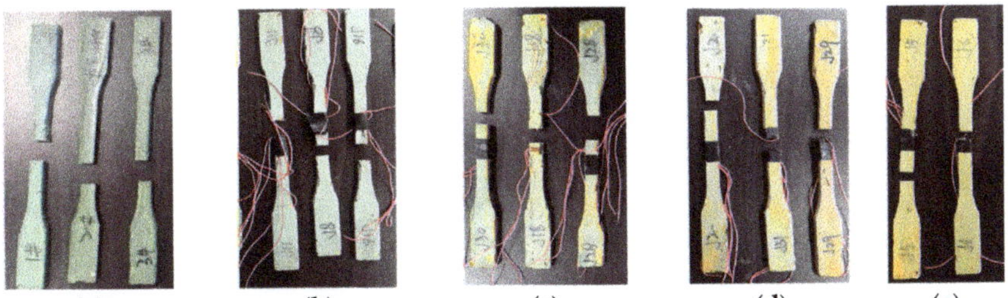

Figure 3. Failure mode of the ERA specimens. (**a**) 0 cycles, (**b**) 30 cycles, (**c**) 60 cycles, (**d**) 90 cycles, (**e**) 120 cycles.

3.1.2. Ultimate Load and Peak Strain

The ultimate load and peak surface strain of the specimens are shown in Table 2. Table 2 shows that the ultimate load and surface strain of the ERA specimens initially increased and then decreased with increased F–T/W–D cycles. The largest values were in the 30 F–T/W–D cycles, which were around 44.7% and 147.1% higher than those of the control group, respectively. However, the smallest values were in the120 F–T/W–D cycles, which were around 18.1% and 1.1% lower than those of the control group, respectively. Analysis showed that a positive effect of the F–T/W–D cycles on the bearing capacity of the ERA occurred at the initial stage of the F–T/W–D cycles. However, the positive effect gradually weakened and became a negative impact with increased F–T/W–D cycles.

Table 2. Ultimate load and Peak strain of the ERA.

Specimen Number	Ultimate Load (kN)	Mean Value (kN)/SD	Peak Strain (με)	Mean Value (με)/SD
C0–1	0.94		3433	
C0–2	0.99	0.94/±0.04	4457	3930/±418
C0–2	0.90		3899	
FT/DW30–1	1.29		9629	
FT/DW30–2	1.37	1.36/±0.05	9827	9712/±84
FT/DW30–3	1.42		9680	
FT/DW60–1	1.31		4064	
FT/DW60–2	1.41	1.32/±0.07	6982	4880/±1498
FT/DW60–3	1.24		3596	
FT/DW90–1	1.19		5310	
FT/DW90–2	1.22	1.23/±0.04	5412	5344/±48
FT/DW90–3	1.28		5311	
FT/DW120–1	0.66		4082	
FT/DW120–2	0.93	0.77/±0.12	4215	3885/±377
FT/DW120–3	0.72		3357	

3.2. Bonded Joints

3.2.1. Failure Modes

The bonded joints had six common failure modes (types A to F) [45]. Type A was CFRP delamination with fibers peeling of the CFRP surface, which may be due to the too thick ERA layer and the sufficient bond strength, which enabled easy peeling of the CFRP surface [46,47]. Type B was fracture of CFRP plate. In this failure mode, CFRP plate fractured transversely. Type C was the peeling failure of the bonded interface between the CFRP plate and ERA layer with no fiber adhering onto the surface of the ERA and no damage of the ERA layer. Type D was peeling failure of the bonded interface between the ERA layer and steel plate with no damage to the ERA layer. Type E was the damage of the ERA layer with ERA fractured or ERA debris adhered onto surface of the CFRP plate or the steel plate. Type F was steel-plate yielding with steel plate fractured or largely deformed.

The typical failure modes of the bonded joints are shown in Figure 4. Experimental results showed that the F–W/W–D cycles had little effect on the failure mode, and most specimens underwent type A + C failure. The exceptions were specimens FT30–DW30–1, FT30–DW30–2, and FT60–DW60–3, which underwent type A + C + E failure as shown in Table 3. Furthermore, most specimens underwent bonded-interface failure in the shorter side of the CFRP plate. Therefore, the experimentally observed diagonal failure could be explained as it provided the weakest path to failure. This finding was due the fact that once interface cracks appeared, the axis of the specimen deviated laterally and was no longer in axis tension [48]. The strength of the bonded joints primarily depended on the fiber interlayer strength and the bond strength of the interface between the CFRP plate and ERA [32]. The strength of the ERA played a secondary role for the ultimate bearing capacity of the bonded joints. However, the strength of the steel plate had no effect on the strength of the bonded joints [31,32]. The bond-interface failure characteristics that the carbon fiber attached onto the surfaces of the ERA layer gradually decreased with increased F–T/W–D cycles, and the area of type C increased.

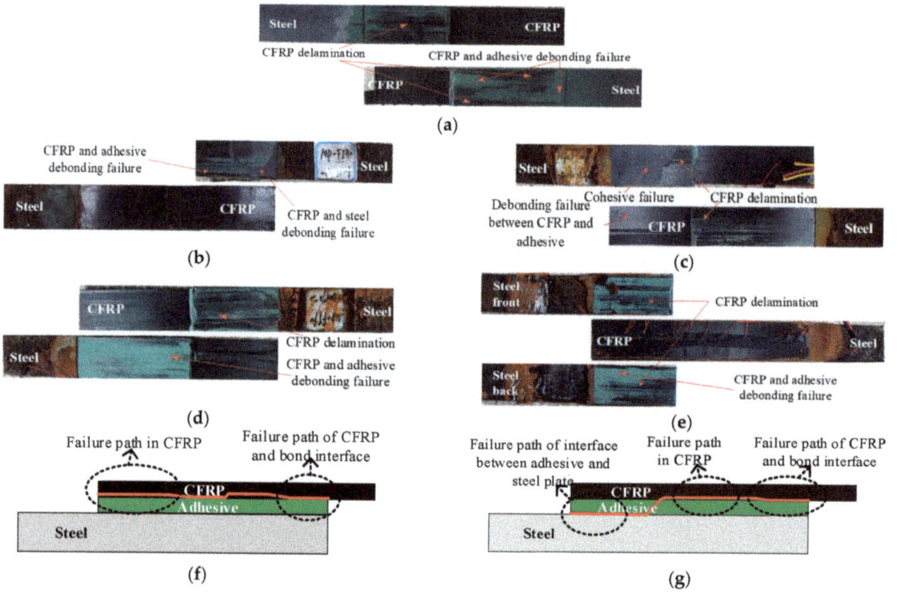

Figure 4. Typical failure modes of the specimens. (**a**) Type A + C (the control group). (**b**) A + C (the F–T/W–D30–1). (**c**) A + C + E (the F–T/W–D60–3). (**d**) A + C (the F–T/W–D90–3). (**e**) A + C (the F–T/W–D120–1). (**f**) Main failure mode I of the bonded joints. (**g**) Main failure mode II of the bonded joints.

Table 3. Ultimate load, interfacial peak strain and main failure modes of the bonded joints.

Specimen No.	Ultimate Load (kN)	Mean Value (kN)/SD	Peak Strain (με)	Mean Value (με)/SD	Failure Mode
C–0–1	75.0	74.0/±1.4	2252	2338/±86	A + C
C–0–2	73.0		2424		A + C
FT30–DW30–1	68.8	67.0/±4.5	2769	2487/±327	A + C + E
FT30–DW30–2	60.9		2028		A + C + E
FT30–DW30–3	71.4		2664		A + C
FT60–DW60–1	40.0	45.0/±4.5	1614	1741/±101	A + C
FT60–DW60–2	44.2		1862		A + C
FT60–DW60–3	50.9		1746		A + C + E
FT90–DW90–1	44.1	45.4/±1.4	2025	2286/±261	A + C
FT90–DW90–2	30.2		366		A + C
FT90–DW90–3	46.7		2546		A + C
FT120–DW120–1	28.2	31.1/±3.0	1894	2119/±169	A + C
FT120–DW120–2	48.3		2161		A + C
FT120–DW120–3	34.1		2302		A + C

A = CFRP delamination; C = Interfacial failure between the CFRP and ERA layer; E = ERA fractured (ERA debris adhered onto surface of the CFRP plate or the steel plate). The acronyms A + C, A + C + E are represented by a mixed damage mode.

3.2.2. Ultimate Load and Peak Strain

The ultimate load of the bonded joints and the peak surface strain of CFRP plate were recorded during the tensile process of all specimens, as shown in Table 3. Compared with the control group, the ultimate loads of the F–T/W–D specimens gradually decreased. Under the 120 F–T/W–D cycles, the ultimate load decreased by 58.0%. Further comparison showed the influence of the F–T/W–D cycles on the ultimate load of the ERA specimens and the bonded joints were consistent.

4. Discussion

4.1. ERA Specimen

4.1.1. Stress–Strain Relationship Curve

Figure 5 shows the stress–strain curve of the ERA under each F–T/W–D cycles. The stress–strain relationship curve of the control group showed significant linearity. With increased F–T/W–D cycles, the stress–strain curve of the ERA specimens presented certain nonlinearity, which became particularly obvious after the 60 F–T/W–D cycles.

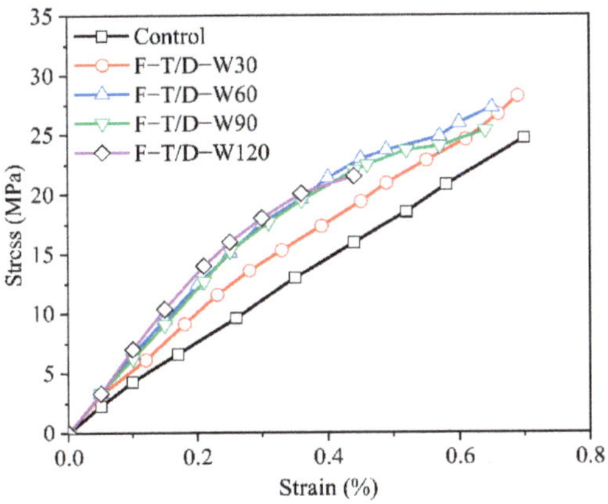

Figure 5. The stress–strain relationship curves of the ERA specimens.

4.1.2. Strength and Elastic Modulus of the ERA

Tensile strength is expressed by the ratio of the ultimate load to the minimum cross–section area of the ERA specimens. The elastic modulus is taken as the maximum slope of the stress–strain relationship curve [49]. From 0, 30, 60, 90, to 120 F–T/W–D cycles, the tensile strength of the F–T/W–D specimens initially increased and then decreased, as shown in the Figure 6a. The elastic modulus initially increased and then unchanged basically, as shown in Figure 6b. This result indicated that the post–curing of ERA played a dominant in the initial stage of F–T/W–D exposure. And as the post–curing was gradually completed, the negative temperature in the F–T environment interrupted the curing process, resulting in a nearly constant elastic modulus [50,51]. Compared with the control specimen, the tensile strength increased around 14.3% under the 30 F–T/W–D cycles and decreased by around 12.9% under the 120 F–T/W–D cycles. The elastic modulus increased by 41.7% and 46.7% under 30 and 120 F–T/D-W cycles, respectively.

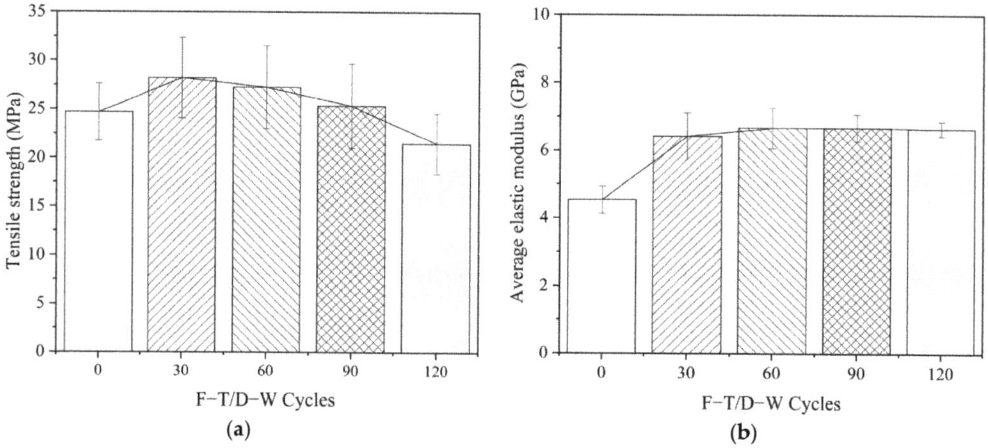

Figure 6. Mechanical property of the ERA. (**a**) Tensile strength, (**b**) Elastic modulus.

The F–T/W–D specimens underwent a certain number of F–T cycles, and the water penetrated the interior of the ERA layer. Consequently, a curing reaction occurred and the tensile strength of the ERA specimens improved. With further freezing action, cracks occured in the interior of the ERA specimens owing to water expansion. Chloride ions in the sodium chloride solution entered the ERA–layer interior through cracks and underwent hydrolysis under the action of the W–D cycle, resulting in a sudden drop in tensile strength of the ERA [36]. Analysis results showed that the presence of salt solution in the F–T environment accelerated the degradation of the ERA's mechanical properties.

4.2. Bonded Joints

4.2.1. Surface Strain Distribution

By collecting the surface strain of the CFRP plate under different loaded stages, the relationship curve between the axial strain distribution of the CFRP plate surface and different loads was obtained. Then the adhesive failure process of the specimens during the tensile process was analyzed.

Figure 7 shows the axial strain distributions along the CFRP plate surface. As shown in Figure 7, for the control group and the F–W/W–D specimens, the strain distributions on the CFRP plate surface were basically similar. The peak strains of the CFRP plate always appeared at the interval of the two steel plates (a strain gauge was attached near here). Further analysis of the strain distribution on the CFRP plate surface showed that the strain decreased sharply from the peak strain point to the adjacent measuring points. Then, the strain changed little toward the load end, and the strain value was close to zero.

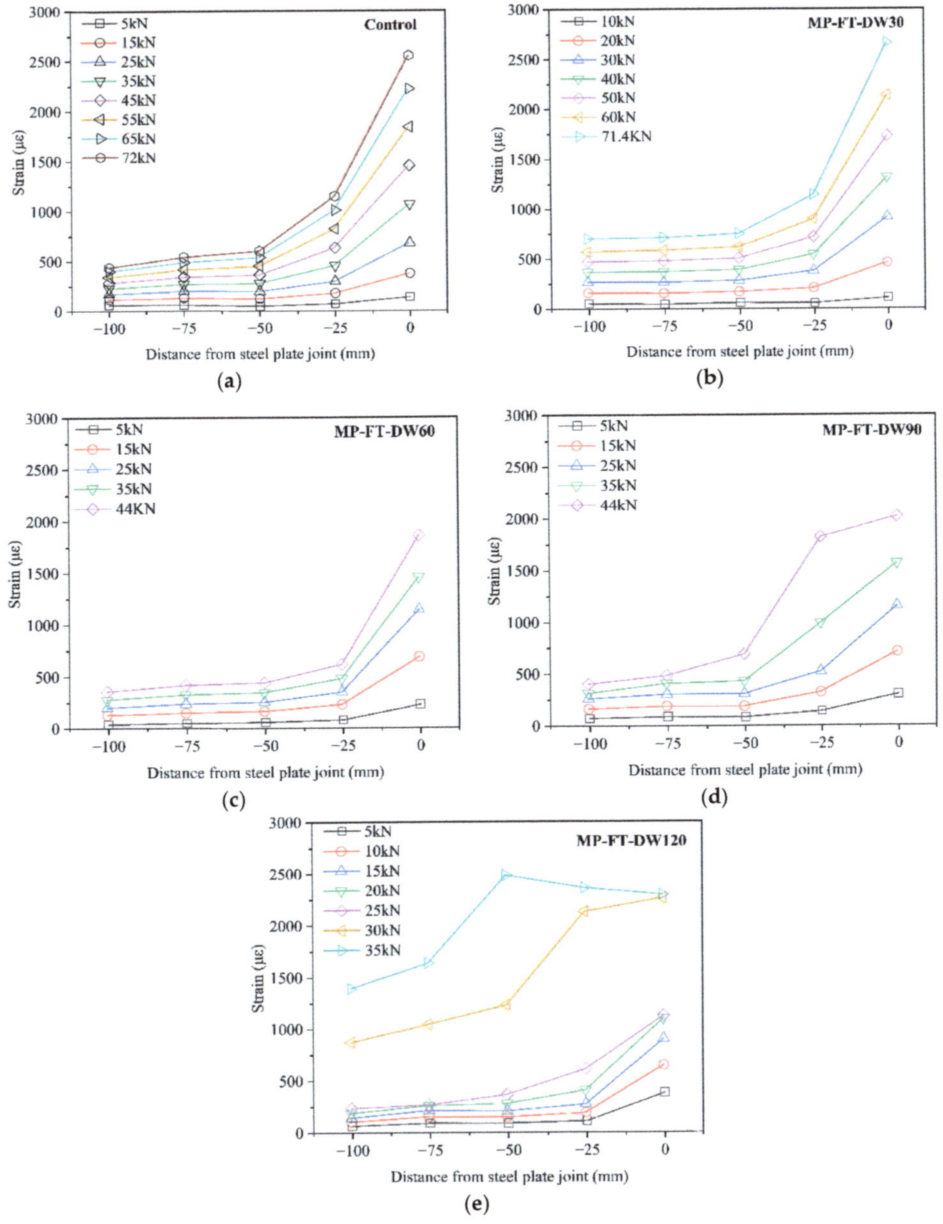

Figure 7. Strain distributions of the CFRP plate surfaces. (**a**) control group. (**b**) 30 F–T/W–D cycles. (**c**) 60 F–T/W–D cycles. (**d**) 90 F–T/W–D cycles. (**e**) 120 F–T/W–D cycles.

The strain concentration area of the bond interface was located near the interval of the two steel plates, and the concentration area was small. With increased F–T/W–D cycles, when the specimen approached the ultimate load, the CFRP surface strain near the interval reached the peak value, and then the value slightly decreased with increased F–T/W–D cycles, particularly after 60 F–T/W–D cycles.

4.2.2. Bond-Slip Relationship

Bond-slip relationship refers to the relationship between the local shear stress and the corresponding slip along the bonded interface of the bonded joints [50]. The bond–slip relationship can be used as the constitutive relationship of the bond interface and could predict the interface failure process of the steel bridge strengthened with CFRP plate. It can characterize the local stress and bond-failure process of the bond interface [51].

Interfacial shear stress and slip can be observed by reading the strain attached onto the surfaces of the CFRP plate by using Equations (1) and (2) [50,52]:

$$\tau_{i-1/2} = \frac{\varepsilon_i - \varepsilon_{i-1}}{L_i - L_{i-1}} E_p t_p \quad (1)$$

$$S_{i-\frac{1}{2}} = \frac{(\varepsilon_i + \varepsilon_{i-1})}{4}(L_i - L_{i-1}) + \frac{(\varepsilon_{i-1} + \varepsilon_{i-2})}{2}(L_{i-1} - L_{i-2}) + \sum_{i=3}^{i} \frac{(\varepsilon_{i-2} + \varepsilon_{i-3})}{2}(L_{i-2} - L_{i-3}) \quad (2)$$

where ε_i is the reading of the ith strain gauge counted from the free end of CFRP plate, $\varepsilon_0 = 0$; L_i is the distance of the ith strain gauge from the free end of the CFRP plate, $L_0 = 0$; E_p and t_p are the elastic modulus and thickness of the CFRP plate, respectively; and $\tau_{i-1/2}$ and $S_{i-1/2}$ are the shear stress and slip at the middle point between the ith strain gauge and the i–ith strain gauge.

As shown in Figure 8, the bond–slip curves were obtained by adopting Equations (1) and (2). As shown in Figure 8, the bond-slip curve under the 120 F–T/W–D cycles had an ascending branch and a descending branch of the bonded joints. However, the bond–slip curves of the control specimens and environmental specimens under the 30, 60, and 90 F–T/W–D cycles had no descending branch. The bond–slip curves of all specimens had a similar bi–linear shape [50,53].

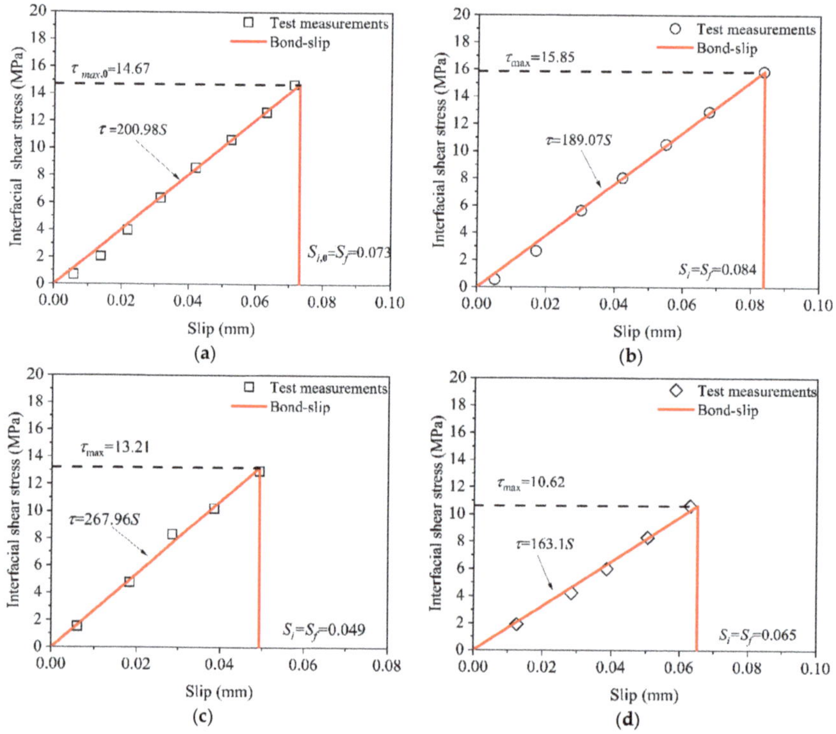

Figure 8. *Cont.*

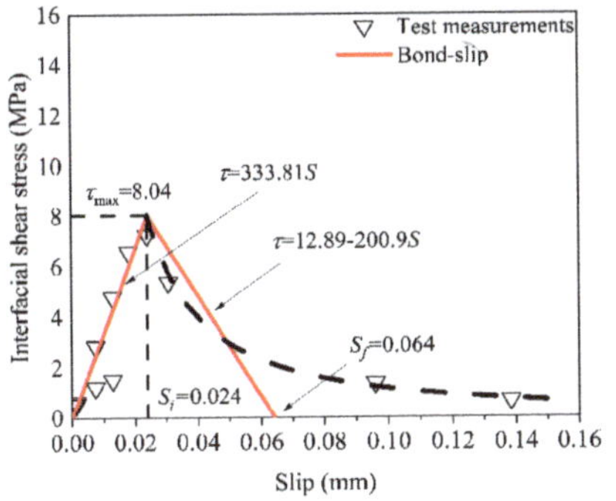

(e)

Figure 8. The bond-slip curves of the bonded joints. (**a**) control group. (**b**) 30 F–T/W–D cycles. (**c**) 60 F–T/W–D cycles. (**d**) 90 F–T/W–D cycles. (**e**) 120 F–T/W–D cycles.

The bond-slip relationship of all specimens can be expressed approximately as Equation (3) [23,50].

$$\begin{cases} \tau = \tau_{\max}\frac{S}{S_1}, S \le S_1 \\ \tau = \tau_{\max}\frac{S_f - S}{S_f - S_1}, S_1 < S \le S_f \\ \tau = 0, S > S_f \end{cases} \tag{3}$$

where τ is the interfacial shear stress; $\tau_{\max}$ is the maximum interfacial shear stress; S is the slip value corresponding to the shear stress τ; S_1 is the slip value corresponding to the maximum interfacial shear stress $\tau_{\max}$; and S_f is maximum slip value. The specific expression is shown in Figure 8.

5. Interfacial Bond-Slip Degradation Relationship

5.1. Interfacial Bond-Slip Parameters

The parameters reflecting the interfacial bond properties of the bonded joints are $\tau_{\max}$, S_1, S_f, and G_f. The parameters $\tau_{\max}$, S_1, and S_f can be obtained from the bond-slip curve in Figure 8. The parameter G_f is approximately expressed as the envelop area of the bond–slip curve [23]. All parameters are listed in Table 4.

Table 4. Bond performance parameters of the CFRP–steel specimen.

Specimen Number	τ_{max} (MPa)	S_1 (mm)	S_f (mm)	G_f (N/mm)	Average Value of Ultimate Load (kN)
MP–FT–DW–0	14.67	0.073	0.073	0.54	74.0
MP–FT–DW–30	15.85	0.084	0.084	0.67	67.0
MP–FT–DW–60	13.21	0.049	0.049	0.32	45.0
MP–FT–DW–90	10.62	0.065	0.065	0.35	45.4
MP–FT–DW–120	8.04	0.024	0.064	0.10	31.1

5.2. Degradation Models of Interfacial Bond-Slip Parameters

The damage factor $D(n)$ was introduced to describe the degradation of the interfacial bond-slip parameters after the F–T/W–D cycles. The damage factor $D(n)$ was determined by Equation (4) [23]:

$$D(n) = 1 - X(n)/X_0 \tag{4}$$

where $X(n)$ is the bond-slip parameters after n F–T/W–D cycles, and X_0 is the parameters value without F–T/W–D cycles.

The values of the parameter S_1 were the same as those of the parameter S_f for the specimens of 30, 60, and 90 F–T/W–D cycles. However, for the specimen of 120 F–T/W–D cycles, the experience time from S_1 to S_f was very short, so the parameter S_f was not discussed in this paper. Figure 9 presentes the damage evolution of the interfacial bond–slip parameters after F–T/W–D cycles. Compared with the control specimen, the damage factors of the parameter $D(\tau_{max})$, $D(S_1)$ and $D(G_f)$ slightly decreased under the 30 F–T/W–D cycles because of the interfacial enhancement owing to the post-curing of the adhesive [23]. However, they increased after 60 F–T/W–D cycles. Equations (5) and (6) were used to fit the variation trend of the parameters with increased F–T/W–D cycles.

$$D(n) = an \tag{5}$$

$$D(n) = bn^2 + cn \tag{6}$$

where n is the number of the F–T/W–D cycle; and a, b and c are the fitting coefficients of Equations (5) and (6) as shown in Table 5. The expressions of the interfacial bond-slip parameters were obtained by substituting Equations (5) and (6) into Equation (4):

$$X(n) = X_0(1 - D(n)) = \begin{cases} X_0(1-an) & X = \tau_{max} \\ X_0[1-(bn^2+cn)] & X = S_1, G_f \end{cases} \tag{7}$$

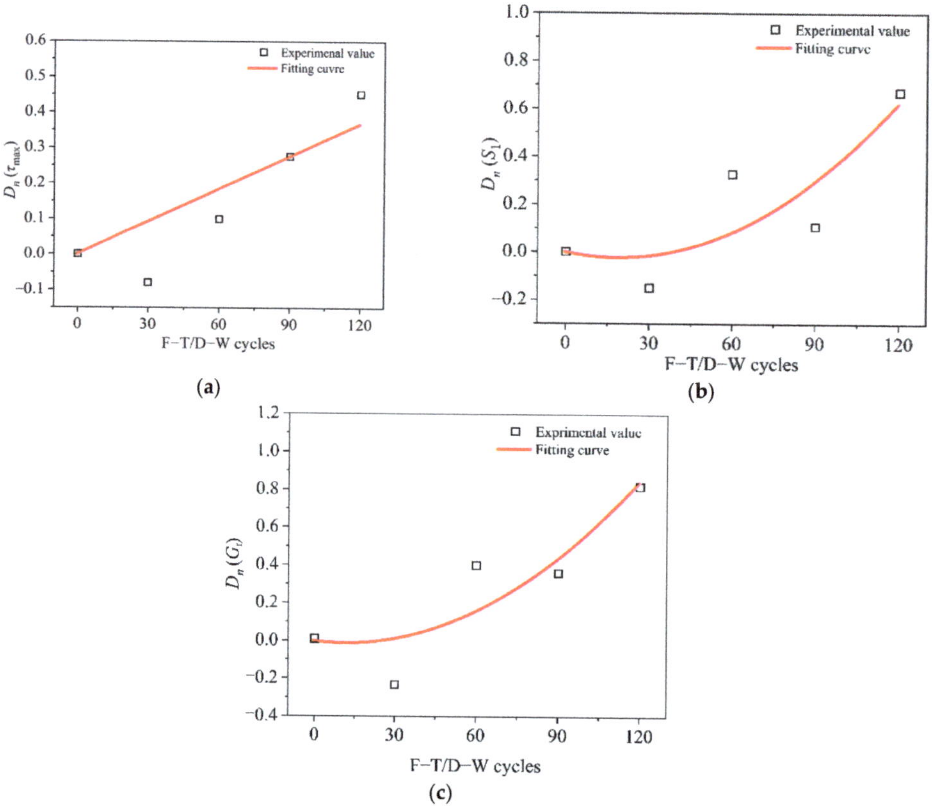

Figure 9. Degradation evolution of the interfacial bond-slip parameters. (**a**) Maximum shear stress τ_{max}, (**b**) Relative slip S_1. (**c**) Interfacial fracture energy G_f.

Table 5. The fitting coefficients a, b and c.

Bond-Slip Parameter	a	b	c
τ_{max}	0.0031	/	/
S_1	/	6.3414×10^{-5}	−0.0024
G_f	/	7.3647×10^{-5}	−0.0018

Formula (7) can express the relationship between the parameter τ_{max}, S_1, and G_f of the bonded joints and the number of F–T/W–D cycles.

5.3. Bond-Slip Degradation Relationship

The interfacial bond-slip degradation relationship was obtained by substituting Equation (7) into Equation (2), yielding Equation (8):

$$\begin{cases} \tau = \frac{\tau_{max,0}(1-an)}{S_0[1-(bn^2+cn)]}S, S \leq S_0\left[1-\left(bn^2+cn\right)\right] \\ \tau = 0, S > S_0\left[1-\left(bn^2+cn\right)\right] \end{cases} \quad (8)$$

where $\tau_{max,0}$ and S_0 are the maximum interfacial shear stress and corresponding slip value without F–T/W–D cycles, respectively. Figure 10 shows the comparisons of the predicted and experimental bond–slip degradation relationship. The predicted degeneration relationship well agreed with experimental ones, except for specimens under the 60 F–T/W–D cycle, which had 23.1% maximum deviation between the maximum experimental shear stress value and the maximum predicted.

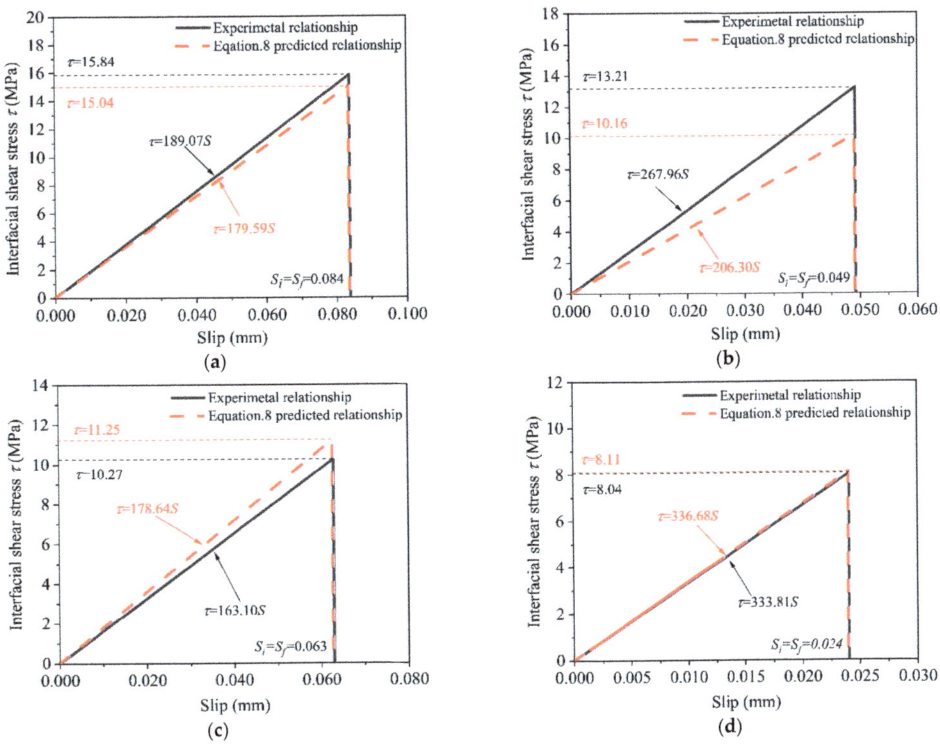

Figure 10. Comparisons of the experimental and predicted bond-slip degradation curves. (**a**) 30 F–T/W–D cycles. (**b**) 60 F–T/W–D cycles. (**c**) 90 F–T/W–D cycles. (**d**) 120 F–T/W–D cycles.

6. Conclusions

Based on the experimental results and analysis of this research, the following conclusions were drawn.

(1) Fracture failure occurred within the area of a small section for all ERA specimens. The hybrid failure modes of fiber peeling on the surface of CFRP plate and bonded interface peeling between CFRP plate and ERA layer primarily occurred in the bonded joints. A few ones had ERA layer failure, and the fracture failure both of the ERA specimens and the bonded joints can be considered to be brittle, indicating that they were unaffected by the F–T/W–D cycles.

(2) With increased F–T/W–D cycles, the ultimate load and tensile strength of the ERA specimens initially increased and then decreased, whereas the elastic modulus basically remained unchanged. For the bonded joints, the ultimate load decreased gradually, and the peak strain of CFRP plate surface appeared at the interval of steel plates, before sharply decreasing toward the loading end.

(3) The triangular double-line model was used to describe the interfacial bond-slip relationship of the bonded joints under the F–T/W–D cycles. The expression of the bond-slip relationship was obtained, and the degradation models of the maximum peak shear stress and the corresponding slip value with F–T/W–D cycles were established based on the fitting. Then, the bond-slip relationship models of the bonded joints were proposed, and the models were verified by comparing with the experimental ones. On this basis, the bond-ship relationship under F–T/W–D cycles was predicted.

Author Contributions: Conceptualization, X.R.; experimental test, L.J., S.C. and Y.Y.; data curation, L.J. and Y.S.; validation, J.H. and X.R.; formal analysis, X.R., Y.S. and Y.Y.; project administration, X.R. and Y.S.; methodology, X.R., Y.S. and Q.L. All authors have read and agreed to the published version of the manuscript.

Funding: This research was funded by the National Natural Science Foundation of China (No. 51408484 and No. 52108246), and China Postdoctoral Science Foundation (No. 2020M673608XB).

Institutional Review Board Statement: Not applicable.

Informed Consent Statement: Not applicable.

Data Availability Statement: The data presented in this study are available on request from the corresponding author.

Conflicts of Interest: The authors declare no conflict of interest.

References

1. Zhang, Q.H.; Cui, C.; Bu, Y.Z.; Li, Q.; Xia, S. State-of-the-Art Review of Fatigue of Steel Bridge in 2019. *J. Civ. Environ. Eng.* **2020**, *42*, 147–158.
2. Ma, J.; Syn, S.; Yang, Q. Edition of China of Highway and Transport. Review on China's Bridge Engineering Research: 2021. *China J. Highw. Transp.* **2021**, *34*, 1–97.
3. Zheng, Y.; Ye, L.P.; Yue, Q.R. Progress in Research on Steel Structures Strengthened with CFRP. *Ind. Constr.* **2005**, *35*, 20–25. [CrossRef]
4. Cheng, L.; Feng, P.; Xu, S.H.; Meng, X.M. The Technology and Research of Steel Structure Strengthened with CFRP. *FRP/CM* **2013**, 58–63. Available online: http://en.cnki.com.cn/Article_en/CJFDTOTAL-BLGF2013Z1014.htm (accessed on 17 July 2022).
5. Dong, S.E.; Ma, M.L.; Wang, Y.H. The Analysis of Parametric and Reinforcement Coefficient about H-Section Steel Beams Reinforced by Prestressed CFRP Plates. *FRP/CM* **2019**, 17–25. Available online: http://en.cnki.com.cn/Article_en/CJFDTotal-BLGF201904003.htm (accessed on 17 July 2022).
6. Liu, Y.; Zwingmann, B.; Schlaich, M. Carbon Fiber Reinforced Polymer for Cable Structures—A Review. *Polymers* **2015**, *7*, 2078. [CrossRef]
7. Peng, H.; Chou, J.X.; Sun, Y.; Zhang, J.R.; Gong, S. Bond Behavior of Near-Surface-Mounted CFRP to Concrete Structure. *China J. Highw. Transp.* **2019**, *32*, 156–166.
8. Agarwal, A.; Foster, S.J.; Hamed, E.; Ng, T.S. Influence of Freeze-thaw Cycling on the Bond Strength of Steel–FRP Lap Joints. *Compos. Part B Eng.* **2014**, *60*, 178–185. [CrossRef]
9. Ren, H.T.; Li, S.; Gao, D.Y. Bond Behaviour of CFRP and Steel under Dry–Wet Cyclic Condition and Loading. *China Civ. Eng. J.* **2009**, *42*, 36–41.

10. Li, J.H.; Deng, J.; Wang, Y.; Guan, J.J.; Zheng, H.Z. Experimental Study of Notched Steel Beams Strengthened with a CFRP Plate Subjected to Overloading Fatigue and Wetting/Drying Cycles. *Compos. Struct.* **2019**, *209*, 634–643. [CrossRef]
11. Borrie, D.; Liu, H.B.; Zhao, X.L.; Singh Raman, R.K.; Bai, Y. Bond Durability of Fatigued CFRP–Steel Double-lap Joints Pre-exposed to Marine Environment. *Eng. Struct.* **2015**, *131*, 799–809. [CrossRef]
12. Li, H.; Xu, S.H.; Wang, Y.D.; Song, C.M. Deterioration of the Bond Behavior between CFRP and Corroded Steel in Wet–Dry Cycling Environment. *Compos. Struct.* **2021**, *272*, 114–202. [CrossRef]
13. Ren, X.; Wang, Y.; Yu, X.; Sun, Y.M.; Liu, Q.F. Experimental study on the interface bond properties of CFRP–plate–steel–plate Joints in freeze–thaw/dry–wet environments. *China J. Highw. Transp.* **2022**, *35*, 247–258.
14. Jiang, S.F.; Cui, E.J.; Wang, J.; Hui, Z.Z. Bond Behavior of CFRP-Concrete Interface in Chlorine Salt Solution with Wet–Dry Cycles. *J. Build. Struct.* **2022**, *43*, 265–274.
15. Cabral–Fonseca, S.; Correia, J.R.; Custódio, J.; Silva, H.M.; Machado, A.M.; Sousa, J. Durability of FRP–Concrete Bonded Joints in Structural Rehabilitation: A Review. *Int. J. Adhes. Adhes.* **2018**, *83*, 153–167. [CrossRef]
16. Toutanji, H.A.; Gomez, W. Durability Characteristics of Concrete Beams Externally Bonded with FRP Composite Sheets. *Cem. Concr. Compos.* **1997**, *19*, 351–358. [CrossRef]
17. Di Tommaso, A.; Neubauer, U.; Pantuso, A.; Rostasy, F.S. Behavior of Adhesively Bonded Concrete–CFRP Joints at Low and High Temperatures. *Mech. Compos. Mater.* **2001**, *37*, 327–338. [CrossRef]
18. Silva, P.; Fernandes, P.; Sena-Cruz, J.; Xavier, J.; Castro, F.; Soares, D.; Carneiro, V. Effects of Different Environmental Conditions on the Mechanical Characteristics of a Structural Epoxy. *Compos. Part B Eng.* **2016**, *88*, 55–63. [CrossRef]
19. Zhang, P.; Shang, J.Q.; Fan, J.J.; Chen, Q.Z.; Zhu, H.; Gao, D.Y.; Sheikh, S.A. Experimental Study on the Bond Behavior of the CFRP Plate-ECC-Concrete Composite Interface under Freeze–Thaw Cycles. *Const. Build. Mater.* **2022**, *316*, 125822. [CrossRef]
20. Wu, B.; Zhang, S.; Liu, F.T. Effects of Freeze–Thaw Cycling on Mechanical Behaviors of AFRP and Epoxy Adhesive. *J. South China Univ. Technol.* **2016**, *44*, 144–150.
21. Moudood, A.; Rahman, A.; Khanlou, H.M.; Hall, W.; Öchsner, A.; Francucci, G. Environmental Effects on the Durability and the Mechanical Performance of Flax Fiber/Bio–Epoxy Composites. *Compos. Part B Eng.* **2019**, *171*, 284–293. [CrossRef]
22. Shi, J.W.; Zhu, H.; Gang, W.; Wu, Z.S. Tensile Behavior of FRP and Hybrid FRP Sheets in Freeze–Thaw Cycling Environments. *Compos. Part B Eng.* **2014**, *60*, 239–247. [CrossRef]
23. Pang, Y.Y.; Wu, G.; Wang, H.T.; Liu, Y. Interfacial Bond–Slip Degradation Relationship between CFRP Plate and Steel Plate under Freeze–Thaw cycles. *Constr. Build. Mater.* **2019**, *214*, 242–253. [CrossRef]
24. Hua, D.; Lin, J.; Zhang, B. Effects of Salt Spray on the Mechanical Properties of Aluminum–Epoxy Adhesive Joints. *J. Adhes. Sci. Technol.* **2013**, *27*, 1580–1589. [CrossRef]
25. Nguyen, T.C.; Bai, Y.; Zhao, X.L.; Ai–Mahaidi, R. Durability of Steel/CFRP Double Strap Joints Exposed to Sea Water, Cyclic Temperature and Humidity. *Compos. Struct.* **2012**, *94*, 1834–1845. [CrossRef]
26. Karbhari, V.M.; Chin, J.W.; Hunston, D.; Benmokrane, B. Durability Gap Analysis for Fiber-Reinforced Polymer Composites in Civil Infrastructure. *J. Compos. Constr.* **2003**, *7*, 238–247. [CrossRef]
27. Xie, J.H.; Lu, Z.Y.; Guo, Y.C.; Huang, Y.H. Durability of CFRP Sheets and Epoxy Resin Exposed to Natural Hygrothermal or Cyclic Wet–Dry Environment. *Polym. Compos.* **2019**, *40*, 553–567. [CrossRef]
28. Yang, Q.; Xian, G.; Karbhari, V.M. Hygrothermal Ageing of an Epoxy Adhesive Used in FRP Strengthening of Concrete. *J. Appl. Polym. Sci.* **2008**, *107*, 2607–2617. [CrossRef]
29. Sousa, J.M.; Correia, J.R.; Cabral-Fonseca, S. Durability of an Epoxy Adhesive Used in Civil Structural Applications. *Constr. Build. Mater.* **2018**, *161*, 618–633. [CrossRef]
30. Heshmati, M.; Haghani, R.; Ai–Emrani, M. Durability of Bonded FRP-to-Steel Joints: Effects of Moisture, De-icing Salt Solution, Temperature and FRP Type. *Compos. Part B Eng.* **2017**, *119*, 153–167. [CrossRef]
31. Xie, W.Z.; Deng, J.; Li, S. Tensile Bearing Capacity of Steel–CFRP strap joints under Dry–Wet Cycles. *Ind. Constr.* **2015**, *45*, 27–30.
32. Dawood, M.; Rizkalla, S. Environmental Durability of a CFRP System for Strengthening Steel Structures. *Constr. Build. Mater.* **2010**, *24*, 1682–1689. [CrossRef]
33. Ren, X.; Sherif, M.M.; Wei, Y.Y.; Lyu, Y.H.; Sun, Y.M.; Ozbulut, E.O. Effect of corrosion on the tensile and fatigue performance of CFRP strand Sheet/Steel double strap joints. *Eng. Struct.* **2022**, *260*, 114240. [CrossRef]
34. Wei, F.F.; Zhu, D.C.; Wang, H.T.; Bian, Z.N. Experimental Study on Bond Behavior of CFRP Plate–Steel Interface in Freeze–Thaw Environment. *J. Southeast Univ.* **2020**, *50*, 803–807.
35. Yang, Y.G.; Silva, M.A.G.; Biscaia, H.; Chastre, C. Bond Durability of CERP Laminates-to-Steel Joints Subjected to Freeze–Thaw. *Compos. Struct.* **2019**, *212*, 243–258. [CrossRef]
36. Heshmati, M.; Haghani, R.; Al–Emrani, M. Durability of CFRP/steel Joints under Cyclic Wet–Dry and Freeze–Thaw Conditions. *Compos. Part B Eng.* **2017**, *126*, 211–226. [CrossRef]
37. Kim, Y.J.; Hossain, M.; Yoshitake, I. Cold Region Durability of a Two-Part Epoxy Adhesive in Double-Lap Shear Joints: Experiment and Model Development. *Constr. Build. Mater.* **2012**, *36*, 295–304. [CrossRef]
38. *GB/T 51355*; Durability Assessment Standard of Existing Concrete Structures. National Standard of the People's Republic of China: Beijing, China, 2019.
39. *GB 50017*; Steel Structure Design Standard. National Standard of the People's Republic of China: Beijing, China, 2017.

40. *GB/T 2567*; Test Methods for Properties of Resin Casting Boby. National Standard of the People's Republic of China: Beijing, China, 2008.
41. *D638-10*; Standard Test Method for Tensile Properties of Plastics. ASTM International: West Conshohocken, PA, USA, 2010.
42. Schnerch, D.; Dawood, D.; RizkAalla, S.; Sumeer, E. Proposed design guidelines for strengthening of steel bridges with FRP materials. *Constr. Build. Mater.* **2007**, *21*, 1001–1010. [CrossRef]
43. Spaggiari, A.; Dragoni, E. Effect of mechanical surface treatment on the static strength of adhesive lap joints. *J. Adhes.* **2013**, *89*, 677–696. [CrossRef]
44. *GB/T 50082*; Standard for Test Method of Long-Term Performance and Durability of Ordinary Concrete. National Standard of the People's Republic of China: Beijing, China, 2009.
45. Chen, T.; Wang, X.; Qi, M. Fatigue improvements of cracked rectangular hollow section steel beams strengthened with CFRP plates. *Thin Walled Struct.* **2018**, *122*, 371–377. [CrossRef]
46. Bocciarelli, M.; Colombi, P.; Fava, G.; Poggi, C. Prediction of debonding strength of tensile steel/CFRP joints using fracture mechanics and stress based criteria. *Eng. Fract. Mech.* **2009**, *76*, 299–313. [CrossRef]
47. Xia, S.H.; Teng, J.G. Behaviour of FRP-to-steel bonded joints. In Proceedings of the International Symposium on Bond Behaviour of FRP in Structures, Hong Kong, China, 7–9 December 2005; International Institute for FRP in Construction: Brussels, Belgium, 2005; pp. 411–418.
48. Narmashiri, K.; Sulong, N.H.R.; Jumaat, M.Z. Failure analysis and structural behaviour of CFRP strengthened steel I-beams. *Constr. Build. Mater.* **2012**, *30*, 1–9. [CrossRef]
49. Li, C.X.; Cao, X.H.; Ke, L.; Luo, N.H.; Guo, L.C. Effect of high temperature on mechanical properties of epoxy adhesive for structural reinforcement. *J. Build. Mater.* **2020**, *23*, 642–649.
50. He, J.; Xian, G.J. Debonding of CFRP-to-steel joints with CFRP delamination. *Compos. Struct.* **2016**, *153*, 12–20. [CrossRef]
51. Wu, C.; Zhao, X.; Duan, W.H.; Al-Mahaidi, R. Bond characteristics between ultra high modulus CFRP laminates and steel. *Thin Wall Struct.* **2012**, *51*, 147–157. [CrossRef]
52. Yu, T.; Fernando, D.; Teng, J.G.; Zhao, X.L. Experimental study on CFRP-to-steel bonded interfaces. *Compos. Part B Eng.* **2012**, *43*, 2279–2289. [CrossRef]
53. Fernando, D.; Yu, T.; Teng, J.G.; Asce, M. Behavior of CFRP Laminates Bonded to a Steel Substrate Using a Ductile Adhesive. *J. Compos. Constr.* **2014**, *18*, 04013040. [CrossRef]

MDPI AG
Grosspeteranlage 5
4052 Basel
Switzerland
Tel.: +41 61 683 77 34

Polymers Editorial Office
E-mail: polymers@mdpi.com
www.mdpi.com/journal/polymers

www.ingramcontent.com/pod-product-compliance
Lightning Source LLC
Chambersburg PA
CBHW061604210926
51850CB00009B/2209
* 9 7 8 3 7 2 5 8 4 9 3 9 0 *